北

Circles of Fantasy:
Convention
in the Plays of
Chikamatsu

Harvard East Asian Monographs
116

ABOVE: Portrait of Chikamatsu.
From Naniwa miyage *(Souvenir from Naniwa, 1738).*
Courtesy of Torigoe Bunzō.

RIGHT: Chikamatsu's Crest.

Circles of Fantasy Convention in the Plays of Chikamatsu

C. Andrew Gerstle

Published by the Council on East Asian Studies, Harvard University, and distributed by Harvard University Press, Cambridge (Massachusetts) and London 1986

Printed in the United States of America
2 4 6 8 10 9 7 5 3

The Council on East Asian Studies at Harvard University publishes a monograph series and, through the Fairbank Center for East Asian Research and the Edwin O. Reischauer Institute of Japanese Studies, administers research projects designed to further scholarly understanding of China, Japan, Korea, Vietnam, Inner Asia, and adjacent areas.

Library of Congress Cataloging-in-Publication Data

Gerstle, C. Andrew, 1951–
Circles of fantasy.

(Harvard East Asian monographs ; 116)
Bibliography: p.
Includes index.
1. Chikamatsu, Monzaemon, 1653–1725—Criticism and interpretation. 2. Jōruri—History and criticism. 3. Chikamatsu, Monzaemon, 1653–1725—Musical settings. 4. Music—Japan—17th century—History and criticism. 5. Music—Japan—18th century—History and criticism. 6. Musical notation. I. Title. II. Series.
PL793.4.Z5G47 1986 895.6'23 85-19071
ISBN 0-674-13172-X

To my parents
C. A. and M. J. Gerstle

Contents

Illustrations

Charts

Acknowledgments

It is a pleasure to record my gratitude to some of the people who helped me prepare this book. Since my research on this subject began at the dissertation level, I am indebted to Donald Shively, Howard Hibbett, Edwin Cranston, and Itasaka Gen of Harvard University for their guidance and assistance. Torigoe Bunzō, Gunji Masakatsu, and Teruoka Yasutaka made my three years at Waseda University both pleasant and rewarding. Special gratitude goes to the octogenarian, Tsuruzawa Jūzō, whose lessons in chanting Jōruri, so kindly given, enriched my understanding of Bunraku music.

While in Japan, I was also fortunate to have had the cooperation of many who are connected with the theater today, particularly, Sasaki Einosuke and his colleagues at the National Theater in Tokyo as well as Ikoma Takami and his associates at the Bunraku Kyōkai in Osaka. Other Japanese scholars who helped me are Yokoyama Tadashi, Yoshinaga Takao, and Tsukuda Ikki. I want especially to thank Fujita Tamao, the scholar-writer, who provided my wife and me comfortable lodging and generous assistance during our two and a half years at his residence. Kikuchi Akira, Matsunaga Kaoru, Hondō Shun, Imanaga Seijirō, and Itaya Tōru helped me collect the illustrations.

Thomas Harper, Laura Fillmore Evans, and the editors at the Council on East Asian Studies at Harvard made many suggestions to improve my manuscript. Further, I wish to thank Mary

Hutchinson for preparing the index and Mary Frewin and K. S. "Bob" Pinkerton for typing and retyping my drafts.

I must also acknowledge my indebtedness to the Rotary Foundation, the Fulbright Commission, and the Edwin O. Reischauer Institute of Japanese Studies at Harvard for the financial support that enabled me to spend three years researching this subject in Japan. Finally, I would like to thank the Faculty of Asian Studies at the Australian National University for providing a grant to travel to Japan in order to collect the illustrations that enhance the book's attractiveness.

Preface

Reading a traditional work of literature from an unfamiliar culture is not an easy task. Fundamental themes pervade all the world's literature, but each tradition has its own means of addressing the common concerns of the human condition. Learning to read and to criticize a work competently demands that the reader try to understand the conventions that artists have either followed or abandoned. We must not attempt to bring the work within our own literary experience. Rather, we must ourselves enter the complexities of the unfamiliar text. To achieve that, the reader must explore the conventions of the art and learn to respond to and appreciate the alien forms and styles. The critic's task is continually to learn new reading techniques for each new genre and style encountered. Too often readers approach this task without considering that the work is of a different tradition based on unfamiliar conventions.

Chikamatsu Monzaemon (1653–1725) is the foremost playwright of popular Japanese drama. Soon after his death, he came to be revered as the "god of writers" for the Jōruri (Bunraku) and Kabuki theaters. Consequently, there is a large amount of Japanese scholarship about his plays; and, thanks to Donald Keene's *Major Plays of Chikamatsu,* for over twenty years we have had good English translations of eleven of his works. Yet, surprisingly little scholarship or criticism in English deals with this major figure or his many plays. This book undertakes to fill part of that lacuna.

Theater, the art of creating effective illusion on the stage, must rely on conventions that provide the audience, as well as the performers and playwright, with a common ground of suspended disbelief. To help Western readers to find this ground from which they can view Chikamatsu's plays, I have attempted to delineate two fundamental conventions of the Jōruri theater. The first is musical form, upon which the play is organized. The second is cyclical movement within the plays, influenced by Buddhist tradition. Ultimately the two conventions are complementary. There are descending and ascending movements of plot that usually begin and end in the same place after a journey through a dark hell of suffering. Music leads the audience along this journey.

Three types of plays are examined. The first is the long *jidaimono* (history or period) play, which most critics do not regard as unified drama; the second is the *shinjū* (love-suicide) play, which appears difficult for non-Japanese readers to appreciate; and the third is the happy-ending *sewamono* (contemporary-life) play, which many critics have condemned as artificial because of its comic ending. Here the focus will be on those conventions common to all three that directly affect narrative style.

For modern Western readers who cannot enjoy Chikamatsu's works on the stage, these fundamental conventions of Jōruri drama are essential to an understanding of the plays. In particular, the musical notation in the texts can help a modern reader or director understand the performing aspects in greater depth. I do not suggest that directors follow exactly the conventions of Jōruri theater, but, were a director or playwright to attempt to adapt a Jōruri play for performance on a Western stage, he or she could make considerable use of the musical notation to detect dramatic methods employed in the original, and possibly incorporate some of them into the production. In hindsight, one can imagine that George Bernard Shaw, who experimented with notation in his plays earlier in this century and wished to develop a system to guide actors' voices, might have found the Jōruri system of notation intriguing. The notation is also a unique guide to the "reading" of Jōruri by past chanters; it lets us see how they interpreted each new text presented to them by playwrights. Their reading determined how the world saw and heard the play. The fundamentals of the musical formulas of Jōruri could provide stimulating ideas for contemporary theater, for Jōruri contains elements similar to realistic dialogue drama, Broadway-style musicals and classical opera.

Although I show how important it is for the musical notation to be included in translations, one question I have not discussed as fully as needed is how most effectively to incorporate the musical notation into translations. Musical notation can assist the translator in imagining how the lines were delivered in performance—their emphasis and rhetorical weight. However, since the word order will be different in a translation, the translator may wish to adjust the notation to better fit the demands of the language of the translation. Where I have inserted notation into the English version I have retained the original order of the notation marks and have tried to place them approximately above the English equivalent of the Japanese. Future translators may wish to experiment with different methods in order to develop a more effective system for reading and stage production.

The critical perspective in this study is primarily from within the tradition. Critical theories imposed from outside often tend to confuse the reading experience rather than to aid it, because they force upon the author hitherto unknown conventions. The ideal reader, of course, should not be limited to the view from within the tradition, but should also draw upon his or her own experience to judge the plays and compare them with other, more familiar works. But criticism must first take into account the conventions of the playwright's theater and the reading conventions of the period.

Many readers of this book will already have an interest in Japan, its literature, and its drama. For those less familiar with Japanese drama, Donald Keene's short book, *Japanese Literature,* particularly the twenty-page chapter on theater, provides an introduction to the subject. A more recent book, *The Voices and Hands of Bunraku,* by Barbara Adachi, will give the reader a sense of the life of Bunraku through pictures, interviews, and descriptions of performances.

Circles of Fantasy:
Convention
in the Plays of
Chikamatsu

. . . all the story of the night told over,
And all their minds transfigured so together,
More witnesseth than fancy's images
And grows to something of great constancy;
But howsoever, strange and admirable.

Shakespeare, A Midsummer Night's Dream

1 · Introduction

Japan has one of the most highly developed and active theatrical traditions in the world. The singular fact that several different kinds of drama have lived on in an unbroken tradition for centuries makes this tradition unique, and fascinating for anyone interested in the development of theater. In contrast to Western theater of the nineteenth and early twentieth century, which eschewed the "theatrical" in favor of the "realistic," traditional Japanese drama is a theater of histrionics and technique, ranging from the subtle art of reducing action to symbolic movement in the Nō to the grandiose, often comically exaggerated, antics of Kabuki. Japanese drama is a drama of action, not words—whether in the fleeting "non-action" of an intensely poetic moment in a Nō play or, at the other extreme, the carefully choreographed sword-fight in Kabuki. Even words themselves are action on the Japanese stage; they are not simply spoken, they are acted. In Nō they are sung, in Bunraku chanted, in Kyōgen mimicked, and in Kabuki declaimed.

One of the most exciting and satisfying aspects of Japanese theater is its multisensuality. Music is important in most performances, but never usurps the stage as in Western opera or Broadway musicals; in Japanese drama it is, however important, only one element in the performance and can seldom stand on its own. Its role is to carry and give rhythm to the action. The language of the play, as well, whether colloquial dialogue or cast in the cadences of classical poetry, is no more than a part of the whole. Unlike European theater, in Japanese drama, verse has never dominated the stage. Costume, makeup, and props also add to the total effect of the performance, and are often spectacular. The masks and robes of Nō are art objects in themselves; the nearly life-size and certainly lifelike puppets are the central focus of that drama; and, in Kabuki, costume and makeup tell us exactly what type of character the actor is playing. No one of these elements ever eclipses the others; each has moments of glory, but remains nonetheless a component.

Another facet of this multisensuality is the organization of a day's program of Japanese theater. In the West, one goes to the theater in the evening for two to three hours. Electric lighting makes this possible. In Japan, even now, traditional theater is performed from around 11:00 a.m. until about 10:30 p.m., the program being divided into two parts. In the not-so-distant past, Japanese theaters began their programs early in the morning and continued until dusk. Troupe managers and playwrights planned their programs to span twelve hours or more.

Early records of Nō performances and particularly the writings of Zeami Motokiyo (1363–1443), the most important figure in the history of Nō, make it clear that a varied and effective program has from the outset been a prime consideration. In Nō this means a group of relatively short, serious Nō plays are interspersed with comic Kyōgen interludes.

Nō plays are narrow in focus and theme. Most center on a single character and a single emotion, such as jealousy, vengeance, or longing. Each is categorized according to character and theme: God, Warrior, Woman, Mad person, Demon. A program of Nō combines these, usually in that order. Between the Nō are Kyōgen light comedies, usually short and farcical but occasionally longer and more serious in theme. The composition of a program of theater is fundamentally different from a typical Western program. Japanese audiences of the Nō (as well as those of Bunraku and Kabuki) expect variety in a performance, a fact not emphasized in Western Nō criticism as plays have generally been treated individually. But the effect of the juxtaposition of plays within a program is as important as the effect of the plays themselves; for the emotions of an audience will be affected by the preceding play and they will view each play in relation to those around it. A Japanese audience must react to a variety of acting and musical styles in the course of a day of theater.

To understand the dramatic conventions of Jōruri (pre-1800 puppet drama)[1] or Kabuki, it is necessary to examine some of the conventions of Nō, the classical model from which later theater evolved. Zeami says, "Music is the spirit of Nō. The discipline of music is essential to Nō theater."[2] In his many treatises, Zeami repeatedly describes the structure of both the individual plays and the day's program in terms of the musical structure of *jo-ha-kyū* (introduction, intensification, and quick conclusion). The terms *jo-ha-kyū* go back to early Bugaku and Gagaku, the court dances and music imported from the continent. This musical concept has since remained central to all Japanese performing arts. It originally referred to tempo: the *jo* is the slowest and introduces the piece or program; the *ha* is more forceful and rhythmical, though not much faster; and the *kyū* is a quick finale. Zeami considered the *ha* the most important portion of the program where the actor "breaks apart the 'orderliness' of the auspicious *jo* and exerts his skill at different roles to the greatest extent in the most detailed manner."[3] To that extent Zeami used the word *ha* in its literal sense of *yaburu,* to "tear" or "break." In the context of theater, *jo-ha-kyū* refers both to musical and to dramatic rhythm, and it is in the *ha* that dramatic climaxes occur. In the *Nōsakusho* (On composing Nō plays, 1423), Zeami describes the writing of a Nō play as a three-stage process: first, "the choice of a suitable source"; second, "the construction of a *jo-ha-kyū* sequence in five movements"; and third, "the gathering of the words (from source), the creation of a melody, and the writing out of the play."[4] The second stage, where the structure is outlined, is the most important, because here the essence of the play must be carefully organized into five sections based on the principle of *jo, ha* (in three sub-sections), and *kyū*. After this, the appropriate musical mood for each section must be determined as well as the melodic patterns.

The *Nōsakusho* states that the dramatic and musical structure must be completely and minutely planned before the author actually writes out the play. The writing of a Nō drama thus demands that the playwright understand the musical principles upon which the theater is founded. In another work, Zeami discusses how playwrights can learn about music.[5] One method is to practice writing plays. That is, one does not study music to learn to write plays, but, rather in the process of writing plays, one learns the principles of music. In Zeami's system, a Nō play is first musical and only then dramatic.

In the *Shudōsho* (The training of actors, 1430), Zeami states that a program performed at a temple or shrine consists of three Nō and two comic Kyōgen, but, in performances for private patrons, as many as ten pieces are often demanded.[6] He considers the ideal to be a five-play program arranged in the order of *jo, ha*$_1$, *ha*$_2$, *ha*$_3$, *kyū,* with the greatest effort of the actors concentrated on the three *ha* plays. Even if the program were arranged differently, the actors should perform in a *jo-ha-kyū* spirit. An eminent scholar of the Nō, Nose Asaji, points out the practical importance of Zeami's emphasis on the *jo-ha-kyū* organization, suggesting that this system places each drama in the most effective position within the program.[7] A play of lighter musical mood and theme is best at the beginning or end, where it can serve to introduce the program or to lift the spirits of the audience after a number of weighty or sad plays. The musical mood, subject, and tempo of the play determine its placement in the sequence. Each play is thus enhanced by its relation to the plays around it. This is in some ways analogous to the organization of Japanese imperial anthologies of poetry, in which poems are arranged in thematic sequences. Each poem is placed within a sequence so that it draws added meaning from the surrounding poems, while also contributing significance to them.[8] Indeed, this skill of arranging disparate elements into a unified whole is a marked characteristic of all Japanese arts, whether graphic, literary, dramatic, or musical.

Zeami himself never states specifically the sequence of types of plays in the program, but the *Hachijō kadensho* (The eight-chapter *Kadensho*), a treatise which directly influenced the development of Jōruri and Kabuki in the seventeenth and eighteenth centuries, outlines precisely the overall plan of a day-long program of Nō. The program outlines six types of plays. The following is a summary translation:

1. Auspicious God play.
2. Warrior Shura play: exorcism of evil spirit.
3. Wig (woman) or *yūgen* (elegant, sad) play: love.

4. Demon play: those suffering after death; purpose is to induce a spiritual awakening (*hosshin*) by showing the suffering of hell.
5. Morality play.
6. Return to auspicious play. Return to spring.[9]

Auspicious plays begin and conclude the program. In contrast, the middle plays generally are tragic. The warrior, woman, and demon dramas all depict suffering in one of the Buddhist samsaric realms in retribution for transgressions in former lives. This pattern of auspicious beginning, journey through the agonies of hell, and return to the auspicious ending, is what I call the cyclical journey or progression of Japanese drama. The quotation above suggests that pathos is the core of the performance. In religious terms, the purpose of the imaginative cyclical journey in which the audience participates is a kind of "awakening." In the terms of Western dramaturgy, "catharsis" might describe the intended effects of such a performance. The Japanese theatrical program, however, does not conclude with tragedy.

Buddhist terms in a treatise on the Nō theater implies that its theatrical framework has been influenced by Buddhism. The popular Buddhist concept of the Six Realms appears in many Japanese works of literature and drama from the introduction of Buddhism in the seventh century onward, and often plays a crucial role in the structure. The Six Realms are beneath the Four Realms of nirvana, thus parts of the illusory world of desire and places of varying degrees of suffering and varying frequencies of transmigration through death and rebirth cycles. They are not totally unlike the circles in Dante's Inferno. However, human existence, too, is considered one of the realms. The fundamental concept concerning these realms is that cyclical movement—death and rebirth—recurs until one achieves "enlightenment," freedom into the state of nirvana. Suffering and hardship can lead to escape from death and rebirth within the Six Realms.

Cyclical or circular movement has always been a central tenet in Buddhism, and many rituals and iconography are organized on circular patterns. In Japan, this concept is fundamental in works of literature and drama such as the *Heike monogatari* (Tales of the Heike), and Nō and Jōruri plays. One manifestation of this idea is the religious pilgrimage, ever-popular throughout Japanese history. The Yamabushi (mountain ascetics), for example, make a rigorous pilgrimage into their mountain retreats, a journey perceived to be through the various realms, during which the traveler experiences death and finally rebirth into nirvana. The fundamental concept is that suffering, or asceticism, can potentially lead one to enlightenment.

Since this cyclical pattern is also at work in Nō and Jōruri (and generally, though to a lesser extent, in Kabuki), then we can postulate that the focus of traditional Japanese theater is pathos and that every dramatic program is in a sense tragic, since catharsis, or in their terms "awakening after the experience of hell," is its aim. The ultimate goal is the audience's experience of rebirth and a "return to spring." The importance of the cyclical journey through suffering to happiness will become clearer in the discussion of individual plays.

Bunraku puppet theater performances even today follow the same pattern of organization that was developed in the Nō. If the program is to be a selection of scenes from different plays, the day's performance usually begins with a festive and auspicious dance (Sambasō). The central sections will be made up of sad scenes, chanted by senior members of the troupe. The program concludes with a happy, lively scene, perhaps a dance. Full-scale, five-act history plays, which themselves fill a whole program, are structured along these same lines. Act I begins auspiciously in a court setting; there is a crisis, and in Acts II, III, and IV its consequences, usually tragic, are worked out; Act V returns to the court and to a peaceful, happy ending. Each act, then, fills a role similar to a single Nō play in a full program; it has a measure of independence and contrasts in mood and tempo with the other acts around it.

This convention of musical and dramatic variation within a five-act play or mixed program makes Jōruri fundamentally different from Western classical drama, whether Greek or Elizabethan, Racine or Ibsen. The West, from the time of Aristotle, has generally endorsed the theory of unity of dramatic mood, though individual playwrights have often departed from the principle. The aesthetic foundation of Jōruri is quite opposed to that of Aristotle—disunity, or more accurately, dissonance, is a necessary precept for a single five-act play. The reader or spectator of a Jōruri play must not expect Aristotelian unities but rather musical variation, an orchestration of disparate moods instead of a single unifying mood. Dramatic unity is not absent; it is of a different nature, a unity of contrasting and complementary styles.

Jōruri differs from other serious Western drama in another crucial aspect: the actors are puppets. Humans have a subordinate role of striving to make the inanimate figures come alive. Although the characters on the Bunraku puppet stage are nothing more than wood and cloth fashioned by human hands, they appear so real and convincing that audiences often weep during the tragic scenes. In fact, when the same play is performed in both Kabuki and Bunraku, one usually notices more tears in the Bunraku performance. One might conclude, ironically, that the puppet drama is the most realistic of the traditional theaters. Realism, however, is a relative term and, although anyone from the West seeing Bunraku for the first time will certainly be struck by the lifelike actions of the puppets, he or she will hardly feel that the total effect is entirely realistic.

The theatrical experience begins before the curtain opens. A drum signals that the performance will soon commence, and its effect is to prepare the feelings of the audience who have just arrived from the outside world of everyday reality. The drum marks the entry into the world of fiction and imagination. Its tempo gradually increases to a frenzied pace, and the curtain is rushed to the side of the stage, pulled across by an "invisible" figure dressed in black. At the same time, on the right a revolving podium suddenly whirls around bringing a chanter and a samisen player, both in formal attire, into full view. The puppets are already on stage but, before the action begins, a black-hooded figure comes out to announce the play and the names of the chanter and samisen player, finishing, as he began, with the cry of "*Tōzai, tōzai,*" the last syllables stretched out until he disappears backstage. *Tō* is east and *zai* is west; the figure announces the performance to all the world. After a moment of silence, the samisen begins a musical prelude that sets the atmosphere for the particular scene to be performed. Then the chanter slowly and solemnly sings the narrative which describes the scene and leads into the story and dialogue. With his words the puppets come to life.

It is difficult to convince anyone who has not seen Bunraku that puppets can seem to be really alive, but with the nearly lifesize puppets, operated by three men, the illusion is inescapable. There is a common pattern to most spectators' initiation into the world of the puppet theater. At first, the puppets appear to command the stage, even though the main puppeteer is in full view, sometimes even wearing colorful formal costume. The spectator is amazed by the three men who work in absolute harmony to achieve the effect of reality. Usually only the senior operator's face is ever in view, a face striking in its motionlessness even during extremely animated or emotional scenes. For some, the presence of this man is an obstacle; for most, he is a source of fascination.

Paul Claudel (1868–1955), the French poet and playwright who saw Bunraku many times during the 1920s when he was ambassador to Japan, described the puppeteers as "bodiless and faceless, clothed in black sheath, their hands and faces veiled in black," and the puppet as "the collective soul of this patch of shadow, of this group of conspirators whose existence one soon forgets."[10] He found the puppets fascinating because they represented accurately on the stage the life of the playwright's words. Having worked with actors performing his own plays, he appreciated the puppets for the total lack of ego that allowed them completely to "become" the characters portrayed. "The living actor, whatever his talents, always disturbs us by admixing a foreign element into the fictitious part he is portraying, something contemporary or commonplace; he is always in *disguise.* The puppet, on the other hand, draws its life and movement from the action and story . . . It is not the actor speaking, it is the word which acts."[11] Two centuries earlier, Chikamatsu had said that imparting emotion to lifeless puppets is the essence of Jōruri. "Descriptive passages like the *michiyuki,* to say nothing of the narrative phrases and dialogue, must be charged with feeling or they will be greeted with scant applause."[12] The words must act or there will be no life on the stage.

After the initiate has gained some familiarity with the magic of the puppets and with the stories of the repertoire, interest often shifts to the chanter, right of the stage, seated on a dais next to the samisen player. From here, the feelings and voices of the characters emanate, often with incredible force. The

Figure 1. Mid-Eighteenth-Century Jōruri Stage. From Tōsei shibai katagi *(Characteristics of modern theater, 1777). Courtesy of Waseda Theater Museum.*

famous performer Takemoto Gidayū (1651–1714) wrote that the chanter must be a "one-man mimic show." He literally acts all the parts, including the presentation of a third-person narrative, and sometimes the histrionics are grandiose. The contemporary chanter Takemoto Tsudayū is well known for his dramatic, often exuberant style during climactic moments. The chanter's skill in imitating in rapid succession a great range of characters—from children to grandmothers, commoners to samurai—is magnificent to observe. This art is so well developed that recitals by chanters with samisen accompaniment but without puppets are quite common. The chanter's task is to bring the words of his text to life on the stage, and his reverence for the text is exemplified in the ritual of bowing to the text and raising it respectfully above his head at the beginning of each performance.

The least ostentatious element in the performance is the samisen; only after continued exposure to Bunraku does one begin to sense its importance in the performance. The samisen player is actually closest to being a director or conductor. During the performance, the samisen controls the tempo and intensity of the action. The samisen's most direct influence is upon the chanter, whom he guides, sometimes prodding, sometimes reining, his performance. And, of course, he expresses in music the action and emotion of the text. To quote Claudel again: "He questions, he is content, he is anxious, he suffers, he desires, he is angry, he is afraid, he reflects, he groans, he weeps, he mocks, he insults, he suspects, he insinuates, he rages, he roars, he caresses. His function is to prepare the audience. He is in himself the whole audience."[13] Further, the music of the samisen delineates the structure of the play, for, with its preludes and cadences, the samisen signals the divisions throughout the performance, and in this way controls its rhythm. This role is of crucial importance in Bunraku because the plays are structured and written according to musical formulas and conventions.

Although all of these discontinuous elements do come together during performance forming a unified whole, the amalgam nonetheless maintains its multidimensional, multisensual facets. Jōruri, or any drama originally composed to be performed on the stage, relies on extraliterary devices to create a satisfying aesthetic experience. These devices are, of course, "extra" only from the perspective of the reader, as opposed to the playgoer. The proper way to experience dramatic art is to watch and listen to it in performance. This is not always possible. Performing traditions may die out, leaving only texts to be read; or national performing traditions, though still alive, may not be readily available to those who live in other parts of the world. Though most drama is composed for performance in a theater, the West has a longstanding tradition of drama as literature. Although playbooks were published in Japan from the seventeenth century onwards, only in the twentieth century have Japanese scholars (under Western influence) begun to deal with theatrical scripts as literature separate from the stage.

Westerners have never questioned whether Japanese plays might be approached as literature to be read. Ezra Pound, William Butler Yeats, and Arthur Waley felt no reservations in dealing with the Nō as dramatic literature, as texts severed from their natural environment, the stage. Yet in Japan, even today, it is most unusual for Nō plays to be discussed solely as works of literature, and most Japanese who do "read" them do so only in order to learn to chant the plays. There has been even less interest in Kabuki texts as serious literature; the scripts of Kabuki plays were not even published until the late nineteenth and early twentieth centuries. Though Jōruri texts were published in full at the time of their first performance, their primary audience was not ordinary readers but amateur chanters. In approaching Jōruri drama as literature, we must consider how to read a text in which extraliterary, particularly musical, elements are a vital part of the experience of that art.

T. S. Eliot in *The Use of Poetry and the Use of Criticism* provides insight into how to appreciate popular drama. His phrase "levels of significance" describes the elements of a play and the ways members of the audience might perceive its action. "For the simplest auditors there is the plot, for the more thoughtful the character and conflict of character, for the more literary the words and phrasing, for the musically sensitive the rhythm, and for auditors of greater sensitiveness and understanding a meaning which reveals itself gradually."[14] He suggests that there are no exact divisions but that different levels will affect the spectator throughout the performance. This concept of "levels of significance" can be applied to Jōruri drama where the audience is offered just such variety in the course of a full program. In the extended history play, the meaning only gradually emerges from the work as a whole. To discern the overall meaning of a history play, all parts must be viewed in relation to the structure of the whole.

An important element in Eliot's thesis is the relationship of the spectator, or the "reader," to the work of art. In recent years the focus of critical discussions of literature in the West has shifted away from the text in isolation to a more balanced view on the interrelationship of the text and the reading experience. One approach is to analyze how the author constructs the narrative to guide the reader's attitudes towards the characters so as to produce a particular response to the events these characters encounter or bring upon themselves in the course of the tale. Wayne Booth in *The Rhetoric of Fiction* uses the phrase "the ordering of intensities" to describe the narrative technique of leading the reader to the mental state necessary to appreciate what the author considers most important in the work.[15] Although Booth speaks primarily of the novel, his approach is more applicable to the performing narrative arts, where rhetoric and delivery, as well as the "rhetoric" of the music, speak directly to the spectators.

One distinctive feature of Jōruri is that many of its characters are not aristocrats but members of the lower classes. Although most history plays deal with problems at the court or governmental level, those who suffer as a consequence are usually not the nobles themselves but their retainers. Moreover, in plays of contemporary life, those who face death in love suicide are never members of the upper classes but the orphans and outcasts of society. The characters' humble station in life does not, however, lessen their "nobility" in the face of death. In fact they achieve stature in dying because their deaths are self-sacrificial, self-imposed for the purpose of a higher motive—to help another. Self-sacrifice is at the core of Japanese conceptions of the truly tragic.

In Jōruri, it is by means of music that the playwright and performers "order the intensities" of the drama, and, since this is true in performance, it must also be considered in reading the play. Without an understanding of the fundamentals of this rhetoric of music, the reader will miss completely a crucial element of the multisensual and multidimensional experience of the play. Although one can never replace the actual experience of viewing the performance, reading with an understanding of the theory of musical organization and of the musical notation will produce an experience much closer to that intended. The "intensities" of Jōruri theater lie in the experience of suffering and death that comes in the middle section of a medley of scenes or in the middle acts of a single five-act play. Music orchestrates, directs, and prepares the audience to follow the cyclical journey down into a dark hell of pain and up again into the light, offering a new and enlightened vision of the world.

La réalité a été si habilement divisée que l'histoire se passe entièrement dans l'imagination et le rêve, sans le support d'aucune matérialité desobligeante. Par d'autres moyens le Joruri *arrive au même résultat que le* Nô.

Reality has been so cleverly divided that the story happens entirely in the imagination and in a dream, without the aid of any disagreeable materiality. With other means, Jōruri achieves the same result as Nō.

Paul Claudel
"Letter to Professor Miyajima"

2 • *Musical Conventions*

PLAYWRIGHT AND CHANTER

Jōruri's origins lie in the oral storytelling tradition, an important part of medieval Japanese popular culture until the development of printing during the late sixteenth and early seventeenth centuries. Storytellers, usually blind minstrels, made their living by traveling about singing their tales to the accompaniment of biwas (lutelike instruments). Around the year 1600, the puppet theater, which also had existed for many centuries as popular entertainment, joined forces with some of these storytellers and Jōruri drama began. However, these performers chose to use the newly imported samisen instead of the ancient biwa, perhaps because of its more lively sound, or perhaps because of its newness. Since these early singers were blind, their art was transmitted orally from master to disciple.

This tradition of oral training remained important even after performers were no longer blind and still is basic to Bunraku today. A major difference between medieval oral storytelling and Jōruri, however, is that Jōruri plays, from the beginning in the early 1600s, were printed as books. The transition from the medieval period (twelfth to sixteenth centuries) to the early Tokugawa period (seventeenth century) was a change of medium and style, not content.[1] The stories are generally from the former age, whether oral epics, Kōwaka-mai dance pieces, Nō plays, or folk tales. In the early seventeenth century, these old stories were given new life in the context of the five or six acts of Jōruri plays, as well as in *kanazōshi* (books simply written in native phonetic script without Chinese characters). Medieval tales (as well as all classical literature), which had been transmitted orally or in manuscript, were now published in the new medium of woodblock print.

The significance of this transition in Japan from oral or manuscript literature to woodblock print might be understood better if we refer to a similar phenomenon in Western history. Marshall McLuhan, in *The Gutenberg Galaxy,* describes a comparable situation in Renaissance Europe,[2] where the revolution in European culture of the fifteenth to seventeenth centuries was also one of medium, that is, from manuscript to the printed book, rather than of content. "During the first two centuries of print, until the end of the seventeenth century, the great body of printed matter was of medieval origin." McLuhan's central point is that homogeneous movable-type print (as opposed to cursive manuscripts) gradually split the oral aspect of the language from the visual; more specifically, in literature, "The divorce of poetry and music was first reflected in the printed page."[3] To support his point, he contrasts medieval reading aloud with post-Renaissance reading in silence. The Elizabethan theater seems to stand as a transitional point in this process; for Shakespeare, it was not essential to publish his plays outside the oral world of the theater, yet in Ben Jonson we see a clear preference for the printed text.

From early in the 1600s Jōruri texts were published for playgoers and readers. In the beginning, the plays contained only the *tayū*'s (chanter's) name. Yokoyama Tadashi has found that, as early as the 1610s, works were published in movable type, but that, by the 1620s, the medium had changed to woodblock print.[4] Most publishing at this time gave up movable type for woodblock, and it was not until the nineteenth century that movable type came back. Japanese culture of the Tokugawa period found its expression through the skilled hands of woodblock carvers, not movable-type printing machines.

The difference of media is significant, because Chinese characters and Japanese syllabary are not carved in homogeneous block style, but are similar to the cursive calligraphy of earlier

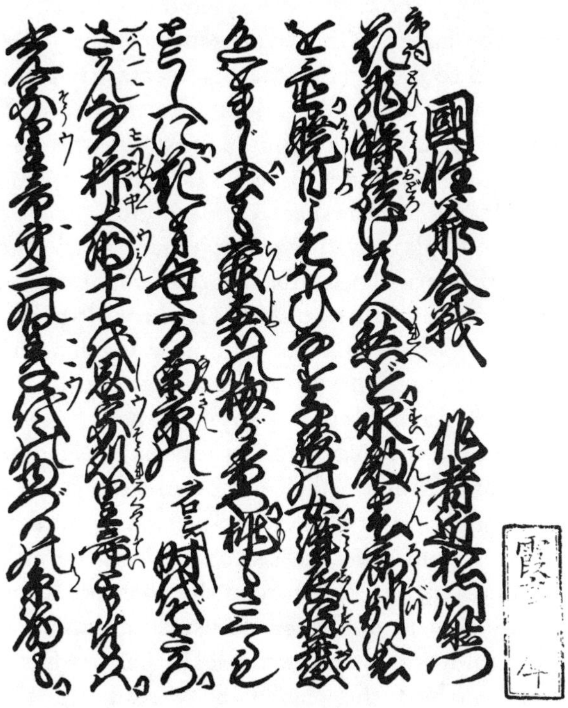

Figure 2. First Page of Kokusen'ya kassen *(The battles of Coxinga, 1715).*

manuscript culture. The use of Chinese characters and phonetic syllabary, though generally standard, does have consistent variety; there are choices of symbols for each sound of the phonetic syllabary, and different Chinese characters for the same word. Perhaps McLuhan, who noted that the divorce of poetry and music followed the development from cursive to standard type, would have said that the woodblock script, with its printed but cursive style, is still closely attuned to the aural aspects of the language. The calligraphic style of Jōruri books is distinctive. The strokes are fat and bold, not beautiful like delicate Heian calligraphy—but, energetic and powerful (see Figure 2). In McLuhan's term they are "tactile" and closer to manuscript than to modern print, which is standard and squarish.

Jōruri texts eventually came to be called *maruhon* (or *shōhon*), which means they contain all the musical notation of the chanter, with none of the secret notation left out. However, until the time of Inoue Harimanojō (d. 1685?) and Uji Kaganojō (1635–1711) in the last quarter of the seventeenth century, the musical notation was held to be a secret treasury transmitted orally from the *tayū* to his assistants and students. By the 1670s the audience's demand for texts with complete musical notation prodded the chanters Inoue Harimanojō in Osaka and Uji Kaganojō in Kyoto, in 1674 and 1678 respectively, to publish collections of acts from different plays with extensive notation.[5] These texts were directed primarily towards amateur chanters. Nishizawa Ippū in *Imamukashi ayatsuri nendaiki* (A chronology of the puppet theater, 1727) wrote that Harimanojō was talented in music and put his efforts into the melodies of the musical scenes of the plays.[6] Ippū also said that people wished to learn Jōruri chanting but found the Jōruri picture books insufficient, so a publisher in Osaka, Inoue Yahei, got permission from Harimanojō to print a collection of scenes complete with the secret notation. From this account we can see that it was the influence of the audience and patrons that forced the chanters to include all the musical notation in published plays. Contemporary readers understood the importance of the musical aspect of Jōruri drama. They expected to hear the play as they read it, because they read with their mouth and ears as well as their eyes.

Once the secret notation was included in publication, it remained customary throughout the Tokugawa period. Generally two types of *maruhon* texts were printed: texts published in large print, initially with eight lines to a page and later with six or seven, were particularly directed at the amateur chanter and were authoritative reprints or first editions;[7] the smaller-print texts (often including illustrations) with many lines, ten or more to a page, were probably directed more toward the "reader" of Jōruri. Nevertheless, from the 1670s on, proper Jōruri texts, no matter how many lines to a page, were all published with the *tayū*'s notation. Therefore in Jōruri, music and poetry were not separated on the printed page—at least not until after the end of the Tokugawa period in the second half of the nineteenth century. Since that time, modern movable-type editions have never contained all of the musical notation because editors have not considered it essential for modern readers. Only recently has there been a revaluation of this attitude due to contemporary research into the relationship between the musical notation and the reading of Jōruri.

The role of the chanter was very important particularly in the formative period during the seventeenth century. Chanters managed the troupes, and theaters were named after them. Furthermore, since the puppets were simple (handled by one operator until the 1730s), the narrative was more important than the puppets. Playwrights may have written or helped write plays for them but, until the 1680s, texts were considered to be the chanters' product, perhaps because the content of the tale (usually from an earlier source) was considered less important than the fashioning of the story into a musical narrative. There was no sense of copyright for the words, only for the music, because music makes the story a Jōruri play, distinct from the earlier source.

The music of Jōruri is different from the music in Western operas, musicals, or movies. In fact, many Western viewers and even scholars of Bunraku have been unappreciative of this sometimes harsh, screeching music. It is not abstract; its main purpose is to express concretely the action or emotion at a particular moment on stage. Therefore, the music is not always pleasing or relaxing, but often quick, jolting, and solemnly slow within a matter of moments. There is, however, a power in the Bunraku samisen perhaps unsurpassed by any other musical instrument. When samisen performers give a demonstration, they usually will not play a lyrical piece in which they show off their "musical" dexterity. Instead they will demonstrate how to make the instrument weep (all the way from a whimper to wailing tears), sing, dance, laugh, fight, and the like. In performance, these musical effects are spectacular, particularly at the climactic moments when almost chopping-like sounds fly from the strings. At such moments an inarticulate cry bursts through the performer's lips, adding a startling effect that could never be noted on a musical score. These techniques can only be learned from imitation of the master.

The singing and recitation of the chanter complement the music of the samisen. The singing is earthy and unrestrained, often filled with vibrant energy that clearly tells the listener that the art's origins lie far from cloistered aristocracy. The emotions are those of the commoners' society, robust, vigorous, and sometimes crude. As with the samisen, the method of learning is rote imitation of the master; there is no improvisation in Bunraku. Although the musical notation in the texts is nothing like a Western musical score, the symbols do mark the general pitch and style of delivery (from unaccompanied recitative to highly melodious singing), as well as rest stops and cadences. Without oral training, however, it is impossible to pick up a text and recreate the music as with a Mozart score. The notation is, nevertheless, of crucial importance because it is the guide to the musical structure.

What is the playwright's role in this theater where musical conventions are so important? Chikamatsu was the first professional Jōruri playwright, and he wrote for a number of

Figure 3. Late Seventeenth-Century Jōruri Stage. From reprint in Seikyoku ruisan *(A history of singing and music, 1847). Puppeteers and chanter are not in view of audience. Courtesy of Torigoe Bunzō.*

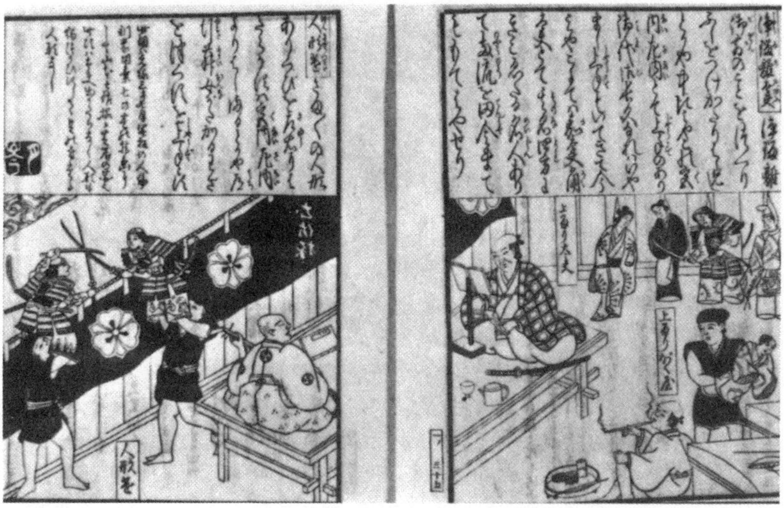

Figure 4. Backstage View of Late Seventeenth-Century Jōruri Stage. From reprint in Seikyoku ruisan *(A history of singing and music, 1847). Courtesy of Torigoe Bunzō.*

chanters during his lifetime. In his famous discussion on the art of Jōruri in the preface to *Naniwa miyage* (Souvenir from Naniwa, 1738), he refers to two chanters for whom he wrote, Uji Kaganojō and Chikugonojō, more commonly known as Takemoto Gidayū. After Gidayū's death Chikamatsu continued writing plays for his successors, the most famous being Takemoto Masatayū (later known as Harima no shōjō, 1691–1744). During his exceptional career of more than forty years, Chikamatsu's relationships with chanters went through many changes. Since Kaganojō was almost twenty years his senior and well established when Chikamatsu was still a fledgling writer, we can suppose that their relationship was strictly master and pupil. Although it is known that Chikamatsu wrote a number of plays in collaboration with Kaganojō, the texts were invariably published with only Kaganojō's name.[8] Kaganojō was one of the last chanters of the age when plays were the sole property of the master *tayū*. Chikamatsu's collaboration with Gidayū, however, was different. They were contemporaries and had been in the theatrical world about the same number of years when they began working together in the mid-1680s. For the first time, Chikamatsu's name appears as the author; however, Gidayū's name appears also at

Figure 5. Portrait of Uji Kaganojō surrounded by his disciples. From reprint in Seikyoku ruisan *(A history of singing and music, 1847). Courtesy of Torigoe Bunzō.*

the end after a postscript which testifies to the accuracy of the musical notation.

Since Chikamatsu had, from around 1700, written almost exclusively for Gidayū, the greatest change in his career occurred with that chanter's death in 1714. Masatayū, although still a young man, became a popular chanter,[9] and it was primarily for him that Chikamatsu wrote many of his most famous plays late in his career, such as *Shinjū ten no Amijima* (Love suicides at Amijima, 1721) and *Onna goroshi abura jigoku* (Woman killer and the hell of oil, 1721). Chikamatsu's relationship with the performers after Gidayū's death was probably quite different from before, because Chikamatsu would have become the experienced member of the troupe. Furthermore, since he was by that time famous outside the theater, he certainly would have had the controlling hand in the creation of his plays.

Jun Shiken compiled a book, *Ongyoku kudensho* (Oral teachings on music, 1771), a collection of the teachings and ideas on Jōruri of his master, Harima no shōjō,[10] full of general comments and brief insights into how a certain passage or character should be presented. These remarks often show remarkable adroitness and perception. Harima no shōjō stressed bringing out the feelings in the text and taught that chanters must always consult the playwright when they do not understand fully the meaning of a passage. This attitude towards the writer had gradually reversed since the days of Kaganojō when the playwright's importance was not even acknowledged; the role of a mature Chikamatsu in the creation of plays was more important than that of the chanter. Harima no shōjō's emphasis on being true to the playwright's text implies that the chanter's art—at

Figure 6. Portrait of Takemoto Gidayū. Courtesy of Tokyo University Komaba Library. Photograph courtesy of Shōgakkan.

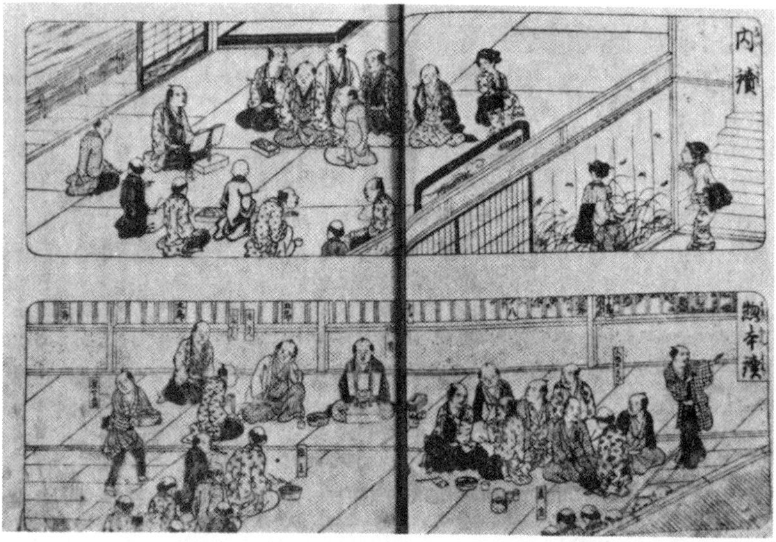

Figure 7. Playwright's First Reading. Nai-yomi *(upper) and* sō-hon'yomi *(below) show the first reading by the playwright of a new play before the troupe. The chanter and samisen player were expected to criticize and suggest changes. From* Shibai gakuya zue *(Illustrations of backstage, 1802). Courtesy of Waseda Theater Museum.*

least after Gidayū's time—came to be evoking the intricacies of the text.

Although it was not the playwright but the chanter who inserted the musical notation in the text, playwrights could hardly have worked without a thorough understanding of the musical conventions. *Shibai gakuya zue* (Illustrations of backstage, 1802), includes important information on the backstage worlds of Kabuki and Jōruri, and gives us some insight into the relationship of chanter and playwright.[11] Two particularly interesting sections present the procedure for preparing Kabuki and Jōruri for the stage. In Jōruri, when a new play was ready, the whole troupe gathered at a teahouse and the producer, chanters, musicians, and puppeteers listened to the *nai-yomi,* or first private reading, by the playwright himself. The chanters, musicians, or puppeteers interrupted the reading if they were dissatisfied with a phrase or the timing. Chanters were also expected to comment on the style and have the playwright alter lines if

Figure 8. Puppeteers' First Practice. Aradate *shows playwright reading text for puppeteers' first practice session. From* Shibai gakuya zue. *Courtesy of Waseda Theater Museum.*

necessary. After this reading, there was a general reading (*sō-hon'yomi*) for the whole troupe and anyone connected with the theater, delivered by the playwright from the chanter's platform. After this, the puppeteers went to their rooms to practice before the playwright, who read the text aloud for them (*aradate*). Next the costumes were prepared (*ningyō-koshirae*). As soon as the chanters and musicians had inserted the notation, each chanter began practicing his own part. After three or four days of practice (*nai-keiko*), the whole troupe marched around town beating a drum (*machibure taikō*), then held a public dress rehearsal (*sō-keiko*) the day before the opening.

It is clear that, at least around the year 1800, the role of the playwright in preparing a new play for the stage was important, because it was through the playwright's oral readings that the performers first encountered the new play. Although we cannot be certain that in Chikamatsu's time, a century earlier, the same procedure for presenting a new play was followed, we can surmise some similarity, largely because Japanese performing arts tend to follow established precedents; and, by the early eighteenth century during Chikamatsu's mature period, most traditions seem to have been fully developed. Furthermore, in the description of this procedure, there are continuous asides by the author that the procedure is very close to the method used in Kabuki. And the method of preparing a play described in *The Actors' Analects,* a book about the Kabuki world of Chikamatsu's lifetime, is comparable to the procedure in use a century later. Since Jōruri has always held more tenaciously to its traditions than Kabuki, we can assume that the observations of Shōkōsai Hanbei in the late eighteenth century reflect the procedures of Chikamatsu's time.

Chikaishi Yasuaki notes an interesting quotation from a preface Chikamatsu wrote late in his life to one of Takeda Izumo's plays.[12] Chikamatsu praises Takeda as a fine playwright and cites his excellence in dialogue and narrative style. He makes an observation that illuminates the playwright's sense of the text: "If you read the words with melody (*fushi*), neither are they too much nor too little for the tongue, and they leave no difficulty for the puppets' movements." Chikaishi concludes that, during rehearsals when the playwright read the text aloud, he read in a voice that contained the variations and melodies chanters used. A further support for this assumption is Chikamatsu's own words in the *Naniwa miyage,* in which he clearly states that his literary style is based on musical formulas: "Thus, while verse is generally written by arranging long and short lines in order, the *jōruri* (*sic*) is basically a musical form, and the length of the lines recited is therefore determined by the melody."[13] Obviously, music was important for the playwright's structuring of his play.

UJI KAGANOJŌ

The playwright in such a procedure was necessarily well practiced as a professional composer of Jōruri. However, when Chikamatsu began writing in his late twenties or early thirties, he knew very little about this drama, veiled as it was in secrecy. Since Chikamatsu was employed to write works to suit Kaganojō's specifications, it is important to know what his chanter's ideas on Jōruri were. Fortunately, Kaganojō wrote on this subject.

Although plays were published from early in the seventeenth century, there were no writings on theory until Kaganojō's preface to *Takenokoshū* (A collection of bamboo shoots) was published in 1683; this is translated in full and annotated in Appendix A.[14] The body of *A Collection of Bamboo Shoots* is a *danmonoshū,* a collection of scenes from Jōruri plays. Kaganojō continued to publish prefaces like this, and generally his ideas are stated in similar terms. From his remarks these collections clearly were published as practice texts for amateur or novice chanters. The selections generally are lyrical *michiyuki* travel songs or other "lyrical-song" scenes rather than more actively dramatic parts, a preference maintained in the later collections of both Kaganojō and Gidayū. The performers considered these lyrical and musical flourishes to be the critical points wherein chanters displayed the heights of their art. There is an incident recorded in *The Actors' Analects* in which Kaganojō explains to his assistants how best to get applause during melodious passages.[15] He says one must not strive for applause but sing the melody evenly, and save one's effort for the lyrical flourishes to make them stand out from the rest. Kaganojō bases his ideas upon this precept of musical variation within a play.

Kaganojō in this preface is defensive and offers an "apologia" for the young Jōruri theater. He traces the art's lineage back to classical drama, naming the Nō as its parent. He states, however, that, since Jōruri does not have teachers, each chanter must develop his own style. Kaganojō's premise is that he himself is the founder of a new tradition.

The discussion of musical styles, which follows his introductory remarks, is the most important section because of its insights into the musical conventions. Musical orchestration within a play is Kaganojō's fundamental principle. He explains, in *A Collection of Bamboo Shoots:* "The basic scheme of the art lies in these four musical styles: the auspicious, the elegant, the amorous, and the tragic." The feelings and musical moods of "each of the four basic musical styles must be chanted differently." Gidayū later used this concept, specifying that each act must have a distinctive style. Kaganojō says that each chanter may create his own melodies and that singing melodies is quite easy, but mastery of the four musical styles takes strict training.

> In an auspicious piece, the sound of the words . . . must be elegant and proper . . . yet they must be delivered with vigor and strength . . .

> For the elegant style the voice should be the same as when chanting auspiciously, but the song should be softer, and the heart should feel as when one is sitting under the blossoms, listening to the koto, flute or *tsuzumi* drum while passing the time of day, forgetting about the journey home. As for the amorous style, the effort of the heart must be more fervent and earnest than for an elegant piece. One must be moved by the lovers, have a yearning heart, and chant with deep concentration . . . Finally, there is the tragic style. For this all the previous sentiments must be forgotten, only a lingering touch of love should remain. Yet the heart must maintain its strength and in its depths should lie the feelings of life's uncertainty, of its transciency.

After this somewhat theoretical section, he discusses discipline and training in the art of chanting, stressing concentration, posture, and enunciation. He emphasizes that, in music, "one must refrain from exerting force into the voice." The chanter's strength must emanate from his heart. Kaganojō's description of the method of chanting as restrained is more reminiscent of Nō than the expressive style normally associated with Jōruri, because he longed for Jōruri to have the tradition and elegance of Nō. His metaphor for Nō drama is an ancient tree—"deep in the mountains near the rocks where the small pines naturally receive the blessing of rain and dew and gradually become covered with moss—such trees are so indescribably magnificent that their beauty and charm penetrate to the depths of one's soul." After this glorification of the classical theater, he reflects on his own position:

> Jōruri too will retain its flavor, if the four musical styles and the enunciation of the syllabary are carefully studied, and if the text is recited to traditional melodies. However, being unable to keep to classical standard, I mix other music, such as popular songs and ballads, into Jōruri, which makes the musical product the same as sickly bamboo or an artificially fashioned tree. Those who have discerning tastes will naturally find this music strange. Since such people are scarce, however, I must follow the whims and fashions of the times, although I myself consider them base.

Though reluctant to waver from classical tradition, he nevertheless adds a note saying that, if a chanter does insert another musical style into Jōruri, he must practice this new melody again and again to make it harmonize with the whole. Despite Kaganojō's protestations against this unorthodox practice, it is this "artificial" mixture of melodies that becomes the essence of Jōruri music.

His use of terms in *A Collection of Bamboo Shoots* shows that Kaganojō was familiar with theoretical treatises on the Nō. Although the authentic texts of Zeami's treatises remained the secret possessions of the traditional Nō schools and were not published till the twentieth century, a spurious edition published during the Tokugawa period influenced the popular theater of the seventeenth and eighteenth centuries. According to a modern introduction to an annotated text of this spurious edition, there are manuscripts as well as printed versions dating from the early seventeenth century.[16] Kaganojō would probably have had access to the edition published in 1665 and reissued a number of times during the Tokugawa period, apparently owing to a wide circulation. Kaganojō, according to his preface to *Shichikushū* (A collection of black bamboo, 1697),[17] tried to enter the traditional world of Nō as a performer but was ultimately unsuccessful because of its closed family system in which only those born or adopted into the tradition can become masters. Given this background, it is easy to understand why he begins by saying the parent of Jōruri is Nō and why he is defensive about Jōruri drama.

Kaganojō believed the status of Jōruri to be plebeian in relation to the patrician Nō, and his idea of making the four musical styles the foundation of Jōruri came directly from the Tokugawa Nō treatise known today as the *Hachijō kadensho.* That text, however, lists five styles;[18] the one Kaganojō omits is *rankyoku,* frantic, fast-paced music. Kaganojō chooses to incorporate only ideas from the *Hachijō kadensho* which concern music, conspicuously omitting any references to dramatic content and organization. It has been generally acknowledged from his own time and throughout the Tokugawa period that Kaganojō was a musical genius. Certainly he made musical theory and technique the basis of his art, and in his prefaces musical considerations preempted all else, including any discussion on the organization of the musical styles within a play. However, he does not give any clear sense of the order of the musical styles, nor how the play is to be structured, and sounds at times as though he is discussing singing lessons for nondramatic music. Nevertheless, his ideas remained influential and are repeatedly referred to in later works on Jōruri. His concept of musical orchestration within a play became a fundamental principle in the tradition.[19]

Since Kaganojō is, of course, addressing his remarks to persons who chant plays, he puts great emphasis on the precept that a chanter must perform the four musical styles, each according to its distinctive characteristics. From the standpoint of musical organization, it is easy to understand his emphasis on variety of musical style, because in music sharp contrasts are necessary devices used by composers and performers to create an aesthetic experience. In Jōruri, musical contrasts are stressed more than Western-style harmonic continuity. This aesthetic consideration was of consequence for playwrights such as Chikamatsu who constructed their plays for chanters in the Jōruri tradition, and this convention is also important for the audience, whether spectator or reader. Although Gidayū made more extensive use of dramatic concepts found in the *Hachijō kadensho,* the principle that each act has a distinctive musical style remained essential for him.

Kaganojō's emphasis on the distinction of musical styles within a play follows directly the Nō tradition in which a theatrical program consists of different kinds of plays. In Jōruri, however, a single five-act play was conceived to fill a whole day of theater, so the musical variety is contained within it. Kaganojō's reverence for Nō must have influenced the young Chikamatsu, who naturally looked to Nō as a source of models. In fact, Chikamatsu refers to more than ninety Nō plays.[20] Moreover, since many of Kaganojō's plays—which are probably collaborations with Chikamatsu—have their sources in the Nō, we can conclude that Chikamatsu learned how to write Jōruri plays while an apprentice to Kaganojō, using Nō plays as models. Nō is not the parent, but it could be called an influential uncle.

In the last section of his short treatise, Kaganojō defines, for the first time in print, the notation found in texts, though he is not the first to have inserted notation in published plays. Yūda Yoshio reviewed pre-Kaganojō Jōruri and found that, from around 1640, notation began appearing, though at first the number and variety were few and simple.[21] In general, there are two kinds of notation—one for lyrical sections, another for the storytelling or dramatic parts. In *michiyuki* lyrical scenes, there are *fushi* or *uta* notations which signify the points when a chanter displayed his singing skills, while the other kinds of notation delineate the end of dramatic or musical units. Initially *kiri* is more prevalent in this role, but, by the 1660s, *sanjū* and *okuri* replace *kiri* in the texts. Tsunoda Ichirō has described the various functions of *kiri:*

1. Change of subject;
   change of scene
2. Movement of characters, exit, entrance
3. Change of atmosphere on stage
4. Climax of battle scenes or violent fights[22]

Often in pre-Chikamatsu Jōruri, new sections of plays begin with set phrases such as *sate mo sono nochi* (well then after that), remnants of the medieval oral storyteller's art. These phrases continued to appear through the early period but faded out in Chikamatsu's time. Chanters used them to make transitions and to serve as musical boundaries between two sections. Gradually it seems that the samisen accompaniment replaced such transitional phrases, because a musical prelude by the samisen came to serve the same function. Even in modern times, however, the tradition of carrying over the last line of the former scene and making it the slow introductory opening passage of a new scene, when there is a change of chanters, has remained customary in performance. This line, the *kaeshi,* simply means a repetition or returning of the last line of the previous scene. Kaganojō's discussion of the "secret" notation, included in full for the first time only a few years earlier in texts published for the general public, was a reluctant though decisive move. Influence of patrons again seemed to prompt that action.

TAKEMOTO GIDAYŪ

Takemoto Gidayū, a former student and later rival of Kaganojō, also wrote some short treatises on Jōruri, and the influence of both his theoretical ideas and dramatic performances proved to be even greater than Kaganojō's in the development of Jōruri theater. By the time Gidayū opened his theater in Osaka in 1684, the practice of including musical notation in published plays was well established. In his preface to a later collection of Jōruri scenes, *Ōmu ga soma* (Parrot Mountain, 1711), Gidayū cautions the reader that, since there are pirated editions of plays in circulation, he should depend only on works of the publisher Yamamoto Kyūemon because, "even if there is only a minute mistake in the musical notation, it makes a great difference. For example, although Chinese ginseng root may look better to the eye because it is smooth, Korean ginseng with its numerous stringy roots is much more salubrious."[23] This comic, earthy metaphor clearly projects his emphasis on the importance of notation for readers. The stringy root is a tactile symbol. Gidayū expects his audience to "feel" their way through the play, using their various senses, including the aural sense. The reader is to hear the music of the play.

It was common practice for the chanter to include a postscript in the published texts, attesting to the accuracy of the notation in the *maruhon.* Some of these postscripts make explicit that the chanter (and playwright) considered the notation to be of significance for readers. Yokoyama Tadashi has examined various types of postscripts in *maruhon* texts. One type contains a sentence that gives us a clear indication of how crucial the chanter considered the notation to be: "The musical notation is important for understanding the structure and central theme of the play, and the feeling of the words."[24] The experience of reading a Jōruri play in Chikamatsu's time was considered by both the composers and readers to be a recreation of the multisensual experience of the stage.

Three years after Gidayū opened his theater in Osaka, he published a collection of scenes from various Jōruri plays. Though the title of the collection has been lost, it is referred to today as *Jōkyō yonen Gidayū danmonoshū* (The 1687 Gidayū collection of Jōruri scenes). The preface is an important treatise, translated in full and annotated in Appendix B. This publication, like Kaganojō's *A Collection of Bamboo Shoots,* was Gidayū's first statement to the world on the status of his art. Both chanters continued to publish collections with prefaces, but generally the principles stated in these first two works remained the basis for their prospective theories. These writings were continually referred to and quoted in later publications on Jōruri throughout the Tokugawa period. Gunji Masakatsu suggests that Gidayū's preface was directed at his rival, Kaganōjo,[25] and the text does seem to contain references to him and his ideas. For Gidayū, Jōruri is a new, independent tradition, which considers "Nō and Kōwaka-mai and other traditional music as foster parents." He uses two metaphors for Jōruri, "the fashionable and confident dandy" and "the bright and delicate blossoms of single-petal Yoshino cherry trees"; both reflect liveliness. In speaking of Nō, he says: "To be sure, the finest trees are the old ones [Nō] deep in the valleys, upon which only a few scattered flowers quietly bloom, but, if one cultivates a young tree [Jōruri] in imitation of these old trees, then its branches will not grow strong; rather they will be stifled, and the blossoms will be feeble and have little charm." Gidayū feels none of the intimidation in relation to "classical" Nō that seems to have haunted Kaganojō. Gidayū emphasizes, also, that art develops not from theory but from diligent training and practice. Since he had a clear sense of being at the formative point in the development of his art, he stresses that melodies must not be restricted to traditional ones, that contemporary songs as well must become a vital source. The artist must "use them to refresh and expand his art. One must open one's ear and mind, because no one school has the secret teachings and traditions which have been handed down from generation to generation." Gidayū is vibrantly positive about the possibilities for Jōruri, and has no chip on his shoulder, in contrast to Kaganojō, who is moving ahead while always feeling that Jōruri is a base and mean art in relation to lofty Nō. Gidayū's ideas are, however, a departure only in theory, not in practice; both chanters mixed new music with traditional melodies.

Gidayū put great faith in an individual's power to achieve artistic distinction through his own fortitude. He was a product of the young Osaka merchant culture, which made self-reliance a prime virtue. Kaganojō, by contrast, lived and worked in Kyoto, where tradition is the way of life. To be sure, Kaganojō does proclaim that art develops from the individual's own heart, but he seems emotionally tied to the past. Gidayū did not live in the confines of tradition, and insists that "each individual must develop his own style and method. As it has been stated in the past, true talent must resemble former and contemporary greatness, and yet be distinctly unique." Zeami, in the *Shikadō* (The way of the highest flower, 1420), refers to a similar saying:

"What one should desire to imitate is skillfulness [of essence]: what one should not imitate is skillfulness [of performance]."[26] Gidayū may be discussing Zeami's distinction between "essence" and "performance," or he may be discussing "individual talent" in relation to "tradition," to use T. S. Eliot's words.[27] Eliot, discussing the relation of a truly new work of art to tradition, says that it will resemble past greatness and conform to the body of "existing monuments," but "to conform merely would be for the new work not really to conform at all; it would not be new, and would not be a work of art." Gidayū's emphasis on individual talent greatly influenced the development of Jōruri drama, because, unlike the other traditional arts, such as Nō and Kabuki which are hereditary, in Jōruri advancement depended, and today depends, almost solely on individual talent and proficiency.

Gidayū, like Kaganojō, took his general ideas on drama from the seventeenth-century Nō treatise, the *Hachijō kadensho,* and adapted them to Jōruri. Kaganojō took primarily musical concepts from the Nō; Gidayū borrowed concepts dealing with dramatic organization and structure. Like Zeami, Gidayū has a good sense of how to give the audience a full and varied theatrical experience. His attitude is practical: "The ability to entertain without boring one's audience should be considered the secret tradition of Jōruri." The most important part of his treatise, the "Outline of the Structure of a Jōruri Play," describes in general terms the make-up of an entertaining five-act play to fill a day's program of theater. There are clear similarities to the program outlined in the *Hachijō kadensho* quoted in Chapter 1. The following is a summary translation.

Act I — Love: entice feelings of audience through pleasant melodies and appearances. "From the opening lines to the end of the prelude is the Sambasō section of the program . . . Since the creation of pleasant appearances should be the primary objective of this act, while your heart must be strong, the words should be light, and your spirit gentle." Silk fabric.

Act II — Shura: (warriors and battles) light mood, vigorous rhythms, and fast beats. "You should chant with speed and force so that the audience listens tensely without loosening their tightened fists." Silk lining.

Act III — Pathos: *kuse.* "When chanting scenes of grief and sorrow, you must not disregard reality. While chanting this act, you must hold the essence of the entire play within your chest." Pattern, color, and embroidery of garment.

Act IV *Michiyuki* (travel song): "In this act the entire play begins to come together as a whole, but since the feelings of the audience are exhausted after the third act, you must be careful not to tire them any further. The melody is the most important aspect of the *michiyuki* scene." Cotton padding.

Act V Conclusion: "The mood of the act is auspicious." Tailoring.

Although this outline contains some dissimilarities from that in the *Hachijō kadensho,* the pattern of beginning with an auspicious (Sambasō and court) setting in a light mood, gradually moving into a darker and more tragic world, and finally returning to harmony is common to both schemes. Gidayū's metaphor for a Jōruri play, a kimono, suggests the multisensuality of the drama: "The first act is the silk fabric; the second, the silk lining; the third, the pattern, color, and embroidery; the fourth, the cotton padding; the final act, then, is the tailoring of the kimono." He is also explicit about towards which sense some acts are directed. In Act I, pleasant appearances and gentle melodies are essential; in Act II, the audience should feel the tension with their fists; Act IV, since melody is most important, is meant primarily for the ear (however, since the *michiyuki* is a dance, the visual element is, of course, essential). McLuhan's notion of relating the tactile and aural senses is certainly evident here as well; the audience (or reader) feels the drama as though caressing a robe of silk. A complete day of theater (here contained in one five-act play) gives the audience a variety of experiences orchestrated to satisfy the many human interests, emotions, and senses. To appreciate Jōruri texts as dramatic literature, the reader must be aware of these conventions.

This structure, which I have called cyclical, is a fundamental convention in traditional Japanese performing arts. One of our most important sources from the seventeenth century on the make-up of Nō, Jōruri, and Kabuki theatrical programs is the diary of a daimyo who was an avid fan of theater and often had plays performed in his domain. This *Matsudaira Yamato no kami nikki* (Diary of Matsudaira, governor of Yamato) covers the years from 1658 to 1695 and includes numerous records of performances.[28] Although the diary entries contain only the titles of the pieces in the program, these titles vividly sketch a picture of a full day of theater in private performance. The program, no matter whether Nō, Jōruri, or Kabuki, always contains variety—some comedy, some acrobatic interludes, some pathos, and some dance. A day of theater invariably begins and ends auspiciously, with the center containing a play on a tragic theme. Kaganojō in his treatise says that, since a Jōruri performance begins with the festive Sambasō dance, the tradition has its roots in the Nō. Although Jōruri was certainly influenced by the Nō, the roots of beginning and ending auspiciously, with tragedy in the middle, go deeper to the fundamentals of the Japanese popular imagination, which was influenced greatly by the cyclical nature of Buddhism.

Gidayū's outline clearly shows his sense of making a day of theater thoroughly entertaining. This is not to imply that the content is merely relaxing, escapist fantasy that skirts sad or tragic themes. Every performance has as its main attraction the depiction of suffering and pathos. In the first act, the audience must be coaxed from the outside world of reality and from the everyday world of work to the world of fantasy and dreams. The audience must then be titillated by exciting action as well as enchanted by pleasant appearances before encountering the central crisis of the play in the third act. The audience is gradually prepared to withstand, absorb, and ultimately profit from the experience of tragedy in the third act by gaining an enlightened perspective on the human condition. After this ordeal the audience is, however, not left stranded and pushed smarting out into the street, but is comforted with pleasant music and dance in the *michiyuki* of the fourth act and finally sent home at the conclusion carrying a vision of a bright future.

In a later work on Jōruri, the *Chikuhō koji* (A history of the Takemoto and Toyotake theaters, 1756),[29] the distribution of roles for each chanter according to his rank in the troupe is outlined.[30] Acts generally are divided into two or three parts: *kuchi* (opening), *naka* (middle), and *kiri* (finale). The following chart lists the various chanters' assignments:

Number 1 chanter: *kuchi* of Act I and *kiri* of Act III
Number 2 chanter: *kiri* of Act IV and *michiyuki*
Number 3 chanter: *kiri* of Act II and *kuchi* of Act III
Number 4 chanter: *kiri* of Act I and *kuchi* of Act IV

Though only a rough guideline, it provides evidence of how the performers and playwrights viewed the play. The structure, based on a system of gradations of difficulty and importance, calls to mind Booth's concept of the "ordering of intensities." The *kiri* of Acts III and IV were considered the most important and the most difficult; even today, only the best chanters will take these parts. In the last part of the Tokugawa period, and today also, little emphasis was put on the opening prelude, but Gidayū felt this was vital to the success or failure of the play. This conventional organization was, of course, well known to the audience, who would naturally alter their expectations for each chanter and for each part of the play. Further, it was within the bounds of this pattern that the playwright produced his work.

Gidayū states that the third act must contain the essence of the entire play, namely pathos. Herein lie the "intensities." The expression of pathos, however, must not disregard reality. Chikamatsu's words recorded in the *Naniwa miyage* (Souvenir of Naniwa, 1738) preface refer to the depiction of pathos in his plays:[31] "*Soregashi ga urei wa mina giri o moppara to su.*" Donald Keene translates this phrase as, "I take pathos to be entirely a matter of restraint."[32] His use of "restraint" for *giri* is an interesting translation. Some Japanese commentators have substituted *dōri,* which means logic or reason. It is, however, not "honor" as Makoto Ueda has translated it in *Literary and Art Theories of Japan.*[33] Yūda Yoshio uses the term *sujimichi,* which also means logic or reason,[34] and explains Chikamatsu's sense of *giri* and *urei* (pathos) in the above passage in terms of how, through the musical and dramatic structure, Chikamatsu gradually constricts the characters' movements until they are caught in an inextricable web. The sense of "restraint," then, could refer to the trapped situation into which the characters have been placed by the structure of the play. If this is close to Chikamatsu's meaning, then the opposite of Ueda's idea that "the plot is subordinate to the character"[35] is true. Rather, through the tightly wrought plot and formal cyclical pattern, the pathos of the characters is revealed. The action and plot of the first two acts set the "restraints" upon the characters in act three; it is then death that frees the shackles and allows the play to reach a happy end. This point will become clearer when we analyze specific plays.

A distinctive aspect of Gidayū's theory is that each act is virtually independent and has a different musical and dramatic mood. What kind of unity is there in Gidayū's Jōruri? Musical orchestration of contrasting acts is a fundamental principle; and Kaganojō's ideas on distinction of musical styles (auspicious, elegant, amorous and tragic) are also certainly an important part of Gidayū's theory, though he does not follow Kaganojō's categories exactly. In Act I, the atmosphere must be pleasant and refined; consequently the setting is usually the court with its pageantry and color. This is, of course, only surface gloss; behind the beautiful façade, a crisis looms. The audience must be sensually coaxed into the world of the stage by pleasant appearances and pleasing melodies, no matter what diabolical schemes are being hatched in the background. In contrast, Act II is comic and fiercely active. The audience must be excited by the prowess of a warrior in action. Usually the setting is not the court but the commoners' world. Act III is always focused on death and sacrifice and so it is, of course, dark and solemn. After this, however, Act IV is again light. Pleasant melodies and appearances in the *michiyuki* must contrast with the bleakness of the previous act and give the audience a chance to relax. Finally, Act V concludes the play by returning to the court setting and to auspicious unity. Gidayū's conception of the unity of a five-act play is different from Aristotle's, or that of a single Nō play; it is a unity of musical harmony among contrasting styles, tempos, and moods.

Each act's composition of varying moods and settings contradicts what is usually considered a principle of unity in Western dramaturgy, demanding new reading techniques. Awareness of the musical and dramatic orchestration provides the key to unity for the reader. Perceiving the play as a cyclical journey through various acts gives a sense of unity to the diverse elements of the play. Although most Jōruri plays end happily, the reader realizes that they are hardly comedy in the simple sense of the word but contain the experience of suffering in hell as the center of a kind of "divine comedy." In a performance the audience is guided by the melodies and rhythms of the chanter and samisen; the reader must seek out this musical guide by feeling his way through the notation, because, in Jōruri, music still clings to the poetry on the printed page.

The drama of pure feeling is no longer in the hands of the playwright: it has been conquered by the musician, after whose enchantments all the verbal arts seem cold and tame.

George Bernard Shaw
"Preface" to Mrs Warren's Profession

3 · *Mosaic Form*

MUSICAL NOTATION

The writings of Uji Kaganojō and Takemoto Gidayū have shown that the plays are best approached as musical dramas. Both chanters outline the meaning of the musical notation, and their postscripts to published plays show that they considered the notation an essential guide to the structure and meaning of the text, and to the life of the words. Without music, the work ceases to be Jōruri. How, then, does notation reveal the narrative structure?

Gidayū describes the organization of Jōruri plays through the following simile: "*Ji* [dramatic song] is like the water; *kotoba* [speech], the current; *fushi* [lyrical song], the quiet depths." In another place he states, "the *serifu* [dialogue] passages are like shallow rapids, the melodic and lyrical *fushi* passages are like quiet pools." Gidayū contrasts two styles within the play: a quick moving and exciting dramatic style and a lyrical style in which action stops, and poetry and music give depth to the play. Yūda Yoshio also describes the basis of Jōruri as the interaction and constant flux between these two styles.[1] One he calls *jigoto* and the other, *fushigoto:*

Jigoto	*Fushigoto*
dramatic or narrative action, dialogue, "shallow rapids"	musical and lyrical song, dance (*michiyuki*), "quiet pool"

This opposition is manifested in two ways. The first refers to the organization within plays into two categories: musical-lyrical and dramatic. Musical sections may be whole scenes such as *michiyuki* or shorter songs within dramatic scenes. On the other hand, dramatic sections contain the action and comprise most of a play. The second manifestation of the opposition occurs in the organization within "dramatic" portions of the work. This aspect is similar to the opposition between recitative and aria in Western opera. In Jōruri, this contrast is between the unadorned, straight-speech style, *kotoba,* and the melodic *ji* or *ji iro* which has musical accompaniment.[2] Before discussing some *jigoto* and *fushigoto* examples, we must first understand the functions of the notation.

Certain notations mark structural divisions in the play. Musically, they signify a cadence at that point, drawing the passage to a close. See Glossaries B and C for descriptions of the notation and the structural units used in the plays.

Yūda Yoshio used the term *yosegi zaiku* to describe the structure of Jōruri drama, and it has also been used earlier by Yokomichi Mario to describe the Nō.[3] The term can be translated as "mosaic." Jōruri plays are composed of many structural parts beginning with the smallest—lines and primary units—and ending with the largest—scenes and acts—all of which form an orchestrated whole. In either tradition the structural parts can be understood better as musical rather than dramatic units. Moreover, these units both are heard in performance and are designated by the musical notation in the published texts.

The fundamental unit in Jōruri's mosaic is what I call the "primary unit" (*tan'i*). A *jigoto* dramatic unit begins with a *ji* or *ji iro* sung line and ends with a musical cadence, designated usually by *fushi* (or by *sanjū* or *okuri* if it ends a scene). A *fushigoto* musical unit begins with *haru fushi.* Ideally, the primary unit is based on the *jo-ha-kyū* musical pattern. It begins very slowly in the *jo,* then moves faster in tempo, as the tension increases in the *ha,* reaching an emotional and dramatic, as well as tempo, peak; a short *kyū* follows, in which the tension and tempo are relaxed during the musical cadence. Between the beginning sung line and the concluding cadence are a variety of chanting styles depending on the context. Usually, there will be a balanced mixture of *ji* or *ji iro* singing and *kotoba* straight speech. The units run from a single line to two or three pages in a modern edition. Though a short one will not be musically complicated, it, nevertheless, will be modeled on the *jo-ha-kyū* pattern. Usually, a short unit contrasts with a long one, which it either precedes or follows. An example of a short primary unit is in Chikamatsu's contemporary and rival Ki no Kaion's (1663–1742) play *Imamiya shinjū marukoshi renrimatsu* (Love suicides at Imamiya near the double pine, 1711).[4] The tale is of a young couple's love suicide. Both work in a large Osaka shop and wish to marry, but Jirobei is still young and unable to support a family. Their master, however, has arranged a marriage for Okisa with a not-so-loyal friend of Jirobei. Jirobei accuses Okisa of being a willing partner in the arrangement. A single line follows a long emotional arialike unit, in which Jirobei is upbraided by Okisa for doubting her love:

ji u *fushi* *haru*
Jirobei is shocked beyond belief and is unable to utter a sound.

This line, designated by the notation as a primary unit, is given added weight, because Jirobei's shame must be expressed in these few words, in contrast to the previous long emotional passage. These units are pieces in the mosaic structure of the play.

Although "mosaic" describes the structure, the form is not static. The units are parts of musical movements. Rhythm in Jōruri is generally based on the *jo-ha-kyū* concept. Each unit has a role, determined by the position of the unit in the specific scene and entire play. In a five-act history play, the reader or audience can expect that Acts III and IV will contain the most critical scenes. Therefore, the primary units in the last half of these acts will be the most intense and difficult. In three-*maki* (scene) *sewamono* (contemporary-life) plays (which were first created from 1703), the climax most often occurs in the last half of the second *maki,* and so the primary units, at that point, will be the most intense and complicated.

The function of *jo-ha-kyū* within a primary unit can be described in two ways. One is that there is a gradual increase in the tempo from a slow, deliberate beginning to a sudden, fast peak, followed by a short, slowing down cadence as the unit comes to an end. In other terms, there is a gradual increase in the tempo from a slow, deliberate beginning to a sudden, fast peak, followed by a short, slowing down cadence as the unit comes to an end. Units vary greatly in length. Early ones in a scene are introductory and bring the audience gradually to a heightened involvement in the intensities of the unfolding drama. The method is to speed up and then slow down the tempo, and to tighten and then relax the tension, bringing the movements to successively higher crescendos until the climax of the particular scene. After the dramatic crisis is passed, the following units serve as a plateau, a relaxing cadence within the whole scene. If in the climax someone dies, then the following units will be an emotional plateau, where the grief of the characters is portrayed. If the drama is successful, the audience's sensibilities will be raised to the point of participating in the emotions following the tragedy. If, on the other hand, this scene is not the climax and is followed by an even more critically dramatic scene, then the audience has been gradually prepared for the intensities to follow, and is allowed a moment of respite before moving on. To be sure, the concept of manipulating an audience through rhetorical and musical devices is not limited to Jōruri or Japanese theater. Any successful drama must bring the audience into a state of intense involvement in the action to give them a satisfying experience. Because, in Jōruri, this rhetorical method of leading the audience through the theatrical experience is based on conventions of music that affect the narrative style and structure of the plays, the notation provides the key to understanding this convention.

JIGOTO UNITS

An examination in some detail of a few examples of "primary units" helps explain how a reader can use the musical notation to understand the narrative techniques. The first two examples are *jigoto* dramatic units; the second two, *fushigoto* song units.

The first is from a Chikamatsu *sewamono* ("contemporary-life" as opposed to five-act *jidaimono* "history") play in three *maki, Shinjū yaiba wa kōri no tsuitachi* (Love suicides in mid-summer with an icy blade, 1709), written for Gidayū. Since the original music (which must be orally transmitted) for virtually all plays from Chikamatsu's time has been lost, we have only the musical notation in the published text as our guide to the narrative style and rhetoric of the language.

The story is the tragic love suicide of a young blacksmith Heibei, orphaned at an early age, and a courtesan, Kokan. Kokan was born into a samurai family, but came to live with her merchant aunt in Osaka after the father became a *rōnin* (masterless samurai). Her aunt fell sick, and Kokan secretly sold herself into prostitution to get money for medicine. In her job as a courtesan she meets Heibei. In the first of three *maki,* the aunt reveals to Heibei that the father (who has no knowledge of Kokan's present status) has been reinstated as a samurai and wishes for Kokan to return home and get married properly. Kokan, however, has no intention of leaving Heibei and says she will die bravely if forced to go. Heibei must get money to buy her contract. He has some customers who have come to pay for a large order, and his boss Riemon welcomes them. After they refuse tea, hot water, and finally tobacco, Riemon realizes that they are from the outcast class and returns their money. This causes a quarrel between Heibei and Riemon, with each in turn succumbing to tears. Riemon finally forgives Heibei on the condition that he change his dissolute ways, but requires him to hold a hot iron as proof of his sincerity. Heibei refuses and, after eighteen years of loyal service, is evicted by Riemon and roughly thrown out by his co-workers.

Maki ii takes place in the pleasure quarter, where we see Kokan for the first time. Kokan, forbidden to meet Heibei, is carefully watched. A *fushigoto* song creates a background atmosphere of bustling festivity. The time is dusk in midsummer during a kite-flying show. The scene is filled with people, colors, and sounds; the air is hot and humid, and emotions, raw and rowdy. Using the confusion brought on by the crowds, Kokan manages to slip away and find Heibei. This encounter is translated below. After she is taken back inside the teahouse, Heibei sneaks in with some ruffians and hides in a closet behind Kokan. They decide to die on the spot if necessary. Kokan's customer is the son of her former nurse, who has come to persuade her to return home. He has already paid her contract, making her a free woman. Kokan is caught between her desire to be with her parents and her love for Heibei. After an emotional scene, she finally agrees to return home, but only as a ploy to enable her and Heibei to have a chance to escape. The final scene of the play is their journey to find a peaceful resting place. They die at dawn after a tortuous journey through the dark night.[5]

ji iro haru
a. (Kokan) "Could he be in this crowd!" she says to herself, pretending to watch the high-flying kites as she walks outside among the
u
throng. She looks to the north and then to the south, catching sight of a man in a light, yellow-striped kimono wearing an oversized wicker hat. "Ah, Heibei!" She reaches out, anxious to find her moonflower in the murky dark. Ahead, he too sees her, though he
u
cannot greet her with his eyes hidden beneath his hat. From their hearts they call to each other. Their eyes look upward, following the
u
kites, and the tips of their fingers reach out; they drift together near the bridge like gently flowing water. They shudder for fear of being
iro *u*
caught, and though close enough to touch, pressed by peering eyes,
ue
they dare not embrace. (Kokan) "Since seeing the letter, my spirit's
u *suete*
dead," is all she can utter before being overcome by tears, which
noru
must be wiped away for fear of drawing notice. Heibei, too, silently weeps beneath his wicker hat.

kotoba
b. (Heibei) "My master is right to disinherit me. I'm not bitter. If you don't go home, you'll suffer in hell for ingratitude to your parents.
ji iro haru
Our future will never look bright. Though I must send you home . . .

u *u*
still how painful it is to part. My love is too strong—I'll plead to your
parents of my pain and suffering. I'll beg them to give their blessing!
fushi
. . . No, it'll never work. We're fated not to be together." He lowers
his head; tears trickle down.

kotoba
c. (Kokan) "I'd finally forgotten them. You've made me weep again.
ji iro haru *u*
How much we should love the stick our parents beat us with. How
u
heinous is my crime against them, who never laid a hand on me and
u *u*
only spoiled me with love! The man who's come to take me home—
we nursed together as babies, and though I can't recall his face, since
he comes from home, I long to see him so. But I've always told them
u
I live in a merchant's house—how could I meet him in this gaudy at-
ue *u* *u*
tire! Thinking of my family, or of you, imagine my agony." She
ji u *u* *chū*
breaks down, her tears like rain. "Memories push me toward the
fushi
country; lingering regrets bear heavily on my heart."

ji iro haru *u* *u*
d. "Determined to die when we next met, I always carry a razor. Here
look at this." Putting her hand inside her sleeve, she pulls out a blade,
with the handle still wrapped, thrusting it into his hands. Then, she
u
grasps his hand saying, "All hail to Amida," and forces the blade to-
ward her stomach. Heibei quickly wrests the razor from her. (Kokan)
"Why stop me? We have as much chance of meeting again as of seeing
u
the mythical Indian udambara blossom! Let me die by your hand!"

fushi
She pleads pitifully in a hushed voice.

kotoba
e. (Heibei) "No, that won't do. In this crowd we won't be able to die in peace. If you're that resolved, let's keep wandering around here. It won't be long till darkness. When it's too dark to recognize faces, we'll walk as far as our legs will carry us. We'll align our bodies proudly in death. Be patient a while longer." He cautions, *ji iro u* "if we are doomed to die, we need not rush." *fushi* Fragile lives, beckoned by the spirit of death.

This long primary unit is followed by a short unit of two or three lines, in which Kokan is discoverd by servants from her teahouse, causing Heibei to escape by scurrying off to the side of the riverbank. These few lines break the intensity of the lovers' passion and replace it with suspenseful tension.

The primary unit translated above is masterfully arranged. Using the notation as our guide (see Glossaries B and C), we can imagine for what effect the playwright and performers were striving. I have divided this single unit into five sections based on the musical notation. The breaks between each come at the *fushi* and *suete* notations. Usually, after the *fushi* musical cadence, a unit will reach a conclusion, but as a rule only if it is followed by a *ji* or *ji iro,* which begins a new unit.[6] The notation *suete* denotes anguished emotions expressed through harsh rhythms in the voice and samisen. Although at the end of a *suete* pattern there is usually a light or incomplete cadence, in most instances it does not mark the end of a primary unit. The play is built upon a foundation of musical units to enable the chanter to pace his performance; he brings each unit to a climax of tension and tempo, which he relaxes during the cadence. Rhetorically, each unit aims to draw the audience further into the excitement and intensities of the drama.

The festive atmosphere of the preceding song carries over to the beginning of this unit. In the crowded quarter, Kokan searches for Heibei. In the distance she spots him disguised in a large wicker hat. They meander through the crowd, gradually coming closer together. The audience, however, knows that Kokan is forbidden to meet the disinherited Heibei. Until this point in the middle of the play, the audience has seen each separately, and so this meeting, their first time together on the stage, must effectively bring the audience into sympathy with their plight. Though they draw close, they cannot embrace. Kokan's first words speak of despair and death, filling this tearful reunion with urgency.

We have no records of how this play was staged, and it has not been performed since Chikamatsu's time. Many of the musical and dramatic possibilities must have been too difficult for the simple puppets and stage of Chikamatsu's time. His technique might today be called cinematic. The first "camera" shot is from a distance: kites fly high over the ground covered with rowdy merrymakers and colorful courtesans. From within the crowd, Kokan emerges into the spotlight. The camera moves gradually inward towards the lovers, until the picture at the end of the first movement is a close-up of the pair alone, amidst the fading crowd. The camera rests on them as they weep. In sections b, c, and d, the crowd disappears altogether, and the close-up fills the stage. Heibei's sad tale prepares the audience for Kokan's *kudoki* (arialike expression of grief), in which she expresses her agony at being torn between love and duty to her parents and love for Heibei. During this central third movement we are coaxed into sympathy for Kokan. Kokan alone commands the stage to portray her grief.

Section d contains quick action in contrast to the preceding still moment of agony. Kokan's attempt to kill herself and Heibei's intervention happen in a flurry. Kokan's despair turns into desperation, and her pent-up wish to die bursts into action. Then, while Heibei explains why he stopped her, the crowd reemerges, as the pair now try to flee from its grasp. Kokan is discovered, and Heibei must run and hide.

The rhetorical technique is effective. Each section builds in intensity to a peak, which is followed by a relaxing cadence. The audience is gradually led into the depths of Kokan's despair. After becoming sympathetic to her plight, however, we are abruptly shocked by the aborted suicide attempt. Finally,

Heibei's rational voice in section e brings the tension and tempo back to a more relaxed level.

Most of the primary unit is in the song style of *ji* or *ji iro,* except for some places where the speech style *kotoba* is used for contrast. Section e is almost all in *kotoba* and so Heibei's rational voice effectively contrasts with Kokan's emotional singing voice. In section c, the variations of voice delivery and musical rhythms, which the notation indicates, are organized to lead the audience into empathy with Kokan's predicament. She begins with a line of *kotoba,* then the voice shifts to *ji iro haru* ("How much we should love . . ."), a singing style in a high pitch, and this continues as she lovingly recalls her family. The *ue,* the highest pitch in Jōruri, is used when she begs Heibei to imagine her agony at being torn between family and lover. Following this musical flourish, she collapses in tears. After this critical moment, the voice changes to a *ji* delivery ("Memories push me . . ."), the most adorned singing style. From the content of this highly emotional line at the end of section c, the audience is led to think that she has decided to return home. This high-strung extreme of her emotions is relaxed a bit in the *fushi* partial cadence, but her monologue continues. Instead of driving Kokan away from Heibei, her feelings of longing for her family thrust her towards him in a desire for peace in death. Both the dramatic and musical possibilities for this violent transition in the midst of Kokan's monologue are magnificent. The cadence has the effect of giving the audience a lingering and suggestive pause to ponder Kokan's mind. Then, while we are wondering where her thoughts will lead, there is a sudden acceleration, as Kokan thrusts herself into the jaws of death. Without notation, the reader is unable to perceive the rhetorical techniques employed, and less able to participate in the drama.

The next example is from Ki no Kaion's play, *Yaoya Oshichi* (The tragedy of Oshichi, 1715), written for Toyotake Echizen no shōjō (1681–1764), a former student and later rival of Gidayū. Although Kaion's plays are not the subject of this book, I have chosen this example from an important work, because it is excellent and because it shows that the use of notation was generally uniform in the works of these two early and influential Jōruri playwrights. This play is also a *sewamono* in three *maki.*

The main characters are Kichisaburō, the son of a *rōnin* and now a young acolyte in a Buddhist temple, and Oshichi, the only daughter of a greengrocer. The pair fell in love when her family took refuge in his temple, while their house was being rebuilt after a fire. The play begins with Oshichi coming with her family to visit the temple and Kichisaburō again. They write out vows of love, but these inadvertently fall into the hands of the villainlike character Buhei, who wants Oshichi for himself. Buhei informs the head priest of Kichisaburō's affair, but the priest calls him a dirty trickster. Oshichi's father, who is also present, asks for proof and the pledges are read aloud. The head priest confesses that he is the culprit, not Kichisaburō. During this confusion, a former retainer of Kichisaburō's father reveals that the father has committed suicide to erase a dishonor and that his last wish was for his son to become a priest. Kichisaburō is caught between duty to his father and the head priest, and love for Oshichi. Kichisaburō is given no choice; Oshichi is dragged away crying.

Maki ii is set on New Year's Eve which should be a joyous time, but Oshichi is miserable. Osugi, Oshichi's maid, is sent on an errand and notices Kichisaburō loitering about in the snow. She sneaks him into the garden and has him hide under the veranda. From there he hears that Oshichi is to be betrothed to Buhei, mainly because Buhei lent them money to rebuild the house after the fire and now demands repayment. Oshichi in the next room has a dream of being attacked and then learns of her fate. She tells her parents they will lose their daughter, if they persist. The mother rationally explains the family's plight. Though the parents did not wish to force her to marry, if she does not accept Buhei, the house must be sold. She tells Oshichi to get Buhei to dislike her, and then get a divorce. Meanwhile, Kichisaburō has been shivering and listening, unable to move. His reaction is the passage translated below. When Oshichi later learns that Kichisaburō had been listening, she goes mad with love and sets fire to the house—hoping to be able to see her lover again.

The play ends at the execution grounds. The parents come to offer Oshichi food; they express their regrets and grief. Oshichi is paraded through the city, to the execution grounds. Kichisaburō pleads to take her place but is refused; he commits suicide in order to guide her in the afterlife, and then she is burned at the stake.

ji iro haru
a. Now a wingclipped fallen bird, Kichisaburō's hands and feet fail to
u
stir. Finally, he struggles out from beneath the veranda, and wipes
chū
his tears on his straw coat, all the while wondering what to do.
haru iro
(Kichisaburō) "Truly there's nothing amiss in her mother's heartfelt
kotoba
counsel. Her pleas leave us no choice. I'm not bitter. Surely, it means
ji chū haru
our love's not to be. How could I have dared to come at all! Heaven's
u
wrath must soon strike me for deceiving my parents and the Abbot.
chū u u
Now the full weight of my transgression crushes my bones. The
fushi
horror of my deeds! What a fool!

chū u u haru ue
b. "Immediately I'll leave and tomorrow take my vows. Never think of
u
me again . . . I have . . . , yes, I've completely forgotten you! It's so
u
easy to say . . . , but could I not just let her know I called? Oh, grant
u
me one last gaze, before bidding farewell." He crawls near the door,
but is startled by his shadow, and quickly jumps back for fear of be-
ing seen from the kitchen. But, in desperation, he approaches again.
suete u
"Why isn't Sugi back here yet!" Overcome with grief, he weeps.

kotoba ji chū u
c. "But I'm wrong again. Oshichi's been betrothed to Buhei. If I steal her,
haru
I'll be an adulterer. If I loiter about, there'll be rumors. How could I
think of touching her nightgown—I won't be needing this disguise

anymore." He discards his cloak and hat; tears fall forming droplets of
haru okuri
ice. Covering his head with his sleeve, he departs.

This is a complicated unit. Following the notation (see glossaries), I have divided the passage into three musical sections with breaks at the *fushi* and *suete* notations. The entire unit begins with an introductory song line *ji iro* and ends with a conclusive cadence, in this case the *okuri,* which signals Kichisaburō's exit. In this little scene, the author and performers want to express Kichisaburō's feelings and reaction to the fatal news, and to bring the audience into sympathy with his dilemma. The notation allows the reader to see the narrative technique; an understanding of the musical conventions enhances the reading experience. The first lines are sung *ji iro haru* with a slow high-pitched melody, and the metaphors suggest our hero's distraught mental state. After the chanter's third-person introduction, his voice becomes Kichisaburō's, and the monologue commences. The opening song style continues until the *iro* ("heartfelt counsel"), which signals a transition from song to *kotoba,* musically unadorned speech. This change occurs as Kichisaburō's mind begins to rationally argue the futility of his case. When he becomes excited, however, the voice shifts to song (*ji*) ("How could I have dared . . ."), and, with the help of the samisen music, the passage builds in intensity to an emotional flourish, as he is frightened by a vision of heaven's wrath descending upon him. The pitch too moves up the scale: *chū, u, haru,* peaking at the emotional climax of section a. With the *fushi* cadence ("What a fool!"), the tension and tempo are relaxed a bit before the chanter continues.

Section b continues in song, but Kichisaburō's thoughts shift direction when he is moved to action. The pitch again rises as he tries to forget Oshichi: *chū, u, u, haru, ue.* Though he struggles to convince himself to leave, his mind wavers indecisively until the section culminates in frustrated anger against the maid Sugi, the lovers' pander, unfortunately away on an errand. The *suete* ("Why isn't Sugi back . . .") musical pattern, through which Kichisaburō's frustrated anger is expressed by harsh rhythms on the samisen and in the voice, is the climax of section b and of the whole unit (because it is the end of the *ha*). In section c, his anger is followed by a return to rationality, and again the delivery is in *kotoba* without musical accompaniment. This is transformed, however, into song (*ji*) as the young acolyte is horrified to imagine himself an adulterer. Finally, in tears he departs, totally dejected. The concluding *haru* pitch returns the passage to the high emotional note that began the unit.

The rhetorical techniques in this passage are easy to see and hear with the help of the musical notation. In section a the audience is made to feel Kichisaburō's horror at his moral transgressions. Then, in section b, we feel his human passion, frustration, and anger, and finally, in section c, we experience his lonely dejection. The fluctuation between rational and emotional thought is effectively conveyed through the contrast between delivery styles of cool speech and hot song. Chanters of that time describe the *ji* delivery as being the most melodious and adorned, and *ji iro* as being less so. We see how finely structured this little passage is by noticing that the chanting pattern is varied and well balanced: emotional *ji iro*—rational *kotoba*—emotional *ji*—rational *kotoba*—emotional *ji*. The shifts back and forth from song to speech effectively highlight the turmoil within Kichisaburō's mind. This contrast is further reinforced by the three-part structure, which contains a series of emotional build-ups followed by relaxations. The audience is guided by the rhetoric of the music into feeling the tortured agony of Kichisaburō, who is caught between his natural desire and the strictures of morality and duty. Without the musical notation, this extra rhetorical element and a consequent dimension of depth are lost.

This example is one primary unit among eighteen in the scene. It is number fifteen toward the end of the *maki* and helps bring the tension to the climax when Oshichi, driven by her frantic desire to see Kichisaburō once again at the temple, goes mad and sets fire to her home.

Each of the units in the scene is similar to the one we have just examined in that there is a gradual increase in the tension and tempo to a peak, followed by a relaxing musical cadence. Yūda has discovered that the *maki* in some of Chikamatsu's *sewamono* plays are organized in five-movement patterns, based on the *jo-ha-kyū* musical form.[7] I have reviewed other Chikamatsu

CHART 1
Musical Structure of *The Tragedy of Oshichi, Maki* ii[a]

Jo A Oshichi's fate of being torn from her lover

1. a. End-of-year preparations for New Year festivities; lovers should be together but cannot.
   b. Oshichi prays the lovers will be joined.
2. The maid Sugi tries to turn Oshichi away from Kichisaburō.
3. Oshichi says their love is sealed in blood.
4. Oshichi's father shows both displeasure with and love for Oshichi.

Ha$_1$ B Kichisaburō loses out to rival Buhei

5. a. Sugi leaves on errand; she notices Kichisaburō in snow.
   b. Kichisaburō says he has come to see Oshichi before he dies.
   c. Sugi conceals him under veranda to wait for her return.
6. Kichisaburō, freezing under veranda, longs for Oshichi's warm body.
7. a. Neighbor pleads for parents to accept Buhei as son-in-law.
   b. Neighbor asks father to persuade Oshichi for family's sake.
8. Triumphant Buhei enters.

Ha$_2$ C Oshichi is persuaded to accept fate

9. Oshichi in next room dreams of being attacked.
10. Servant jokes that she must marry Buhei; he speaks with a stream of fire images.
11. Oshichi threatens to kill herself if they force marriage. Beneath veranda, Kichisaburō writhes in anguish.
12. a. Mother makes long persuasive speech.
    b. Mother's words have fire images; Oshichi will make Kichisaburō suffer in hell.
13. Oshichi resigns herself; mother says she can later get divorced.

Ha$_3$ D Kichisaburō's dilemma

14. Oshichi, confused and flustered, wishes to die.
15. Kichisaburō debates whether to leave or see her once more; he leaves. (a,b,c, three sections).
16. Sugi returns and explains that Kichisaburō has been waiting; Oshichi becomes excited; Sugi blames Oshichi for giving in to parents, and leaves.

Kyū E Oshichi's tragedy

17. Oshichi, now alone, feels she has betrayed Kichisaburō; she weeps while, in next room, parents celebrate marriage.
18. Shift to Nō music as Oshichi goes mad with desire; (musically complicated passage). Oshichi thinks that if house burns, then family will go to the temple again and she will see Kichisaburō. Imagery goes around and around into the fires of Hell.

Note: [a]Capital letters designate five movements of scene (*maki*); arabic numerals mark primary units (*ji* to *fushi*); and small letters mark subdivisions within primary units.

plays that he did not examine, as well as Ki no Kaion's *sewamono* plays, and found that the five-movement pattern is the consistent organizing principle. The above outline of the middle scene of *Oshichi* clearly breaks down into a five-movement pattern. In each of the five movements there is a gradual tightening of the tension and quickening of the tempo to a climax, followed by a relaxation, the same pattern as in a single primary unit.

FUSHIGOTO UNITS

Though *fushigoto* song units constitute (in contrast to *jigoto* units) only a small part of Jōruri plays, they are more crucial than their sparseness implies. Gidayū uses the term, a "quiet pool" (rather than *jigoto* "rapids"), to describe the function of these song passages. Creation of the atmosphere of the scene is one major function of *fushigoto* passages. In all cases, the *fushigoto* passage is a moment when the temporal flow of the drama ceases. In this pause, the audience is offered a new and often revealing perspective on the action and characters. The most prominent example of a *fushigoto* passage is the *michiyuki* travel song, an integral part of most Jōruri plays, which will be discussed in following chapters.

Chikamatsu uses *fushigoto* songs to suggest aspects of the psychology of the characters. In the tragic adultery play, *Horikawa nami no tsuzumi* (Drum of the waves of Horikawa, 1706), he incorporates sections from three Nō plays to heighten the dramatic tensions of the first scene, which culminates in the act of adultery, and to suggest the underlying psychology of the characters. In all three passages, the beating of a Nō drum gradually excites the emotions of the characters and audience to a feverish, lustful level.[8] A section from the Nō play *Matsukaze* (Pining wind) opens the drama on an excited note. The Nō chanting and drum beating in a nearby room are only a practice session, but the story describes exactly the loneliness of Otane. The Nō passages build in intensity to the climactic *kuse* section of the play, when Matsukaze imagines that her long-dead lover has indeed come for her. While listening to the chanting, Otane's sensibilities are excited feverishly to the point when she too believes that her far-away husband Hikokurō has come home, and imagines that a hanging robe is really he. At that moment, however, the chanting and drum stop, leaving the atmosphere on a high plane of emotion. The next *fushigoto* song later in the scene is a drinking-party passage from the Nō play *Rashōmon,* and this time the effect is light and festive as the drum teacher and Otane exchange cups of sake. The final reference is to the demon play *Ominameshi.* The passage is from the fast-paced *kiri* finale in which the Hell of Lust is graphically depicted in song, and accentuated through vigorous drum rhythms. This is used as a background for the attempted rape scene. These three *fushigoto* direct the action and atmosphere, sometimes giving a different perspective on the psychology of a character, sometimes setting the mood, and sometimes commenting on the action.

While *jigoto* primary units begin with the notation *ji* or *ji iro* and end with *fushi* or another notation for a cadence ending, *fushigoto* units usually begin with the notation *haru fushi, fushi,* or *fushi haru* and end with *fushi* (followed by *ji* or *ji iro*). The following example is from Chikamatsu's *Yari no Gonza kasane katabira* (Gonza the lancer, 1717), written for Gidayū's successors. I have supplied the notation to Donald Keene's translation *Gonza the Lancer.*[9] The quotation below includes more than just the *fushigoto* song, so that the "song" unit can be seen in relation to the "dramatic" unit which follows.

This is a *sewamono* play on the theme of adultery, and the main character is Osai, a charming samurai woman of thirty-seven years and mother of two children. Her husband is a tea master away in Edo with the lord of the fief. The man involved in the affair is Gonza, a gallant of twenty-five, a master of martial arts, who aspires to be a tea master as well. Although Gonza has a fiancée, ambition forces him to accept the proposal from Osai to marry her thirteen-year-old daughter Okiku and become the heir of the tea master. The passage below comes just before Osai announces that she wants her daughter to marry the handsome Gonza.

After Gonza is persuaded to marry Okiku, he is asked to return that night to receive the secret teachings necessary to perform a ceremony at the castle. In the meantime, however, a woman comes to ask Osai to be a go-between for the marriage of Gonza and a young woman, Oyuki. This stirs up Osai's jealousy, which she vents in the evening when Gonza comes for the secret teachings. She angrily pulls off Gonza's sash, which was made by Oyuki, and throws it into the garden, where the comic-villain Bannojō happens to be sneaking about planning to enter Osai's room and make love to her. Osai throws her sash at Gonza telling him to wear hers instead. This too is tossed outside and Bannojō, yelling adultery, runs off with both sashes in hand. At this point, Gonza wishes to take his own life, but is persuaded by Osai to flee with her. The final part of the play is their journey and eventual discovery by Osai's husband. Both are killed in the end.

The preceding unit ends with the line:

fushi
More than a mirror, man in this world is the mirror for man.

(*fushigoto* unit)
fushi haru *chū*
The conduct of others serves as our guide to the good and bad in our
own behavior. The artist's brush tells us of the fair women of Kyoto
u *reizen*
and Osaka, and with our mind's eye we can visualize the cherry blos-
soms of Yoshino and Hatsuse: so Osai imagines her daughter as a
haru edo
married woman. "Never let anyone see your hair in morning slumber,
your face when you come from the bath, your hair after washing.
You must be careful when your hair is untidy or disheveled by sleep,
u *fushi*
and even on your pillow your face should be composed.

(*jigoto* unit)
ji iro chū
a. Beauty is granted at birth, but as it says in the *Essays in Idleness,*
haru
well-groomed black hair can make any woman seem lovely; it is her
most precious possession." Osai unravels the secrets of ladylike ap-
pearance and conduct as she strokes her daughter's hair with the fine

teeth of the comb. Under her loving hands, the girl's hair takes on an
fushi
unrecognizably handsome sheen.

kotoba
b. OSAI: There—see what an unusually pretty young lady you've become. If you think I'm not telling the truth, look in the mirror for yourself. But they say a parent's eyes are prejudiced. We need an outsider's judgement. Come here, Man! Sugi, the cook, come quickly! I want you both to look at Okiku's hair.

NARRATOR: The maids come running with cries of "Yes, ma'am!"
ji chū
MAN: My, my, how clever you are, madam! The way you've fixed
u *u* *u*
her bangs and the rest make her naturally pretty face all the prettier.
u
She puts even a girl in a dither. I'd like to hold her naked body tight in my arms and sleep with her.
iro
NARRATOR: Such are Man's praises. Sugi claps her hands.
kotoba
SUGI: So that's it! I've always wondered about something and now at last I understand. Every time I look at my face in the mirror I think how strange it is that no man's ever fallen for a girl with such good features. Now I see I'm such a wallflower, poor thing, only because of the way I do my hair. Excuse me for saying so, but if madam
ji chū
would arrange my hair for two or three days, I'm sure every man in
fushi
the country would fall for me like pampas grass before the autumn wind.

NARRATOR: Her voice is boisterous.

Although there is little action in this hair-combing scene, it is a deep pool in the dramatic flow in which the author suggests much about the workings of Osai's mind. On the surface Osai is giving motherly advice on etiquette to her adolescent daughter; however, the language she uses is definitely erotic. The sequence of references to "hair" reveals the subtleties of the playwright's art. In the following quotation I have arranged the words into a sequence of lines based on the punctuation in the original Japanese text and provided a literal translation.

Tonogo motte no asanegami	Having a husband, your hair in morning slumber,
Yu agari	Or when coming from the bath,
Kao ya araigami	Your face and washed hair—
Hito ni na mise so midaregami.	Never let anyone see. Disheveled hair
Nemidaregami no makura ni mo	Hair disheveled by sleep on your pillow
Negao wa nao mo	And with your sleeping face even more
Tsutsumashiya.	You must be careful.

In Japanese literature the long flowing hair of women has from the *Man'yōshū* (759) onward been a suggestively erotic symbol. Chikamatsu has arranged this sequence to give hints about Osai's psychology. The temporal movement in the passage is from morning to deepest night and from waking to intimate and disheveling sleep. The erotic element is further enhanced by the placing of "*midaregami*" and "*nemidaregami*" in succession. Grammatically, the sentence ending with "*Hito ni na mise so*" comes to a complete stop. The chanter, however, does not pause there, but continues until after the first *midaregami*. This arrangement isolates and intensifies the effect of the image of disheveled hair. Through this technique of isolation and repetition, Osai's words of caution to her young daughter Okiku become explicitly erotic. Today they would be called Freudian in that they reflect her subconscious mind with its excitable desires. Through the *fushigoto* song, the audience is given indirectly, though expressly, a glimpse of the inner workings of Osai's mind, and is gently prepared for the adultery incident. I say gently because only extremely astute members of the audience would be able to follow the intricacies of the song, but they would detect the suggestiveness of the passage due to the repetition of hair and particularly *midaregami,* disheveled hair. Through the medium of song, the erotic language is carefully implanted in the subconscious of the audience to guide its response to later actions.

The entrance of the two maids abruptly alters the atmosphere and introduces a conventional comic *chariba* scene, because the puppet heads for the maids are round and plump, homely, ruddy-cheeked country girls. This contrast of mood is reinforced by the change of delivery styles from song to unadorned speech. However, although the tone is comic, the content is serious. The subtle sexual references in the lyrical passages are expanded into relief through vividly graphic and earthy terms. The maid named Man becomes so excited by the beauty of Okiku's hair that she wishes to sleep naked with the young girl, and Sugi gains new confidence, boasting that, if Osai were to arrange her hair, all the men "would fall for me like pampas grass before the autumn wind." The association of the images of disheveled hair and bending pampas grass is specific enough; both metaphors complement each other and give an erotic undertone to this seemingly innocent scene of arranging the hair of a thirteen-year-old girl. The two primary units deal with the same subject in contrasting ways, and the juxtaposition is effective. In the *fushigoto* passage, suggestive word play and high-sounding references to classical literature are presented in the context of a song; then, in the next passage, earthy and blunt expressions of sexual desire by the two servant girls are put in the context of a standard comic scene. Both modes of expression are purposefully indirect. Erotic imagery is implanted in the audience's mind initially, through pleasant melodies, and later, through laughter. Although the audience may not consciously retain all the implications of the scene, they will, nevertheless, be in a suitable frame of mind to view the unfolding of the tragedy. An analysis of the notation illuminates the rhetorical techniques used by the playwright and performers to present effective drama.

The next example of a *fushigoto* unit is from Chikamatsu's *Meido no hikyaku* (Courier for hell, 1711), written for Gidayū. The following translation is adapted from Keene's *Courier for Hell.*[10] The play is also a *sewamono.* The crisis occurs when

Chūbei, a handsome young courier, steals his customer's money to pay for the contract of his lover, the courtesan Umegawa. In *maki* i, Chūbei's financial troubles are revealed. A country customer is trying to buy Umegawa's contract, and Chūbei does not have the cash to outbid him. Hachiemon, Chūbei's customer, takes pity and loans him money. At the end of the scene, Chūbei sets out to deliver properly a large amount of money, which has just arrived. His desires, however, lead his feet in the opposite direction towards the pleasure quarter.

In *maki* ii, we meet a despondent Umegawa, who sings the song quoted below. After this ends, Hachiemon, flushed with drink, enters the teahouse and tells the courtesans about Chūbei's precarious financial situation. Overhearing this, Chūbei feels slandered, bursts into the room, and, to erase his shame, breaks the seal on the money. With the clang of the gold coins, he and Umegawa are doomed. In the final *maki* iii, the pair flee to Chūbei's father's home in the country. Here they are eventually captured and led off for execution.

The passage cited below is actually incorporated from an earlier play by Chikamatsu, *Sanzesō* (Memorial to Yūgiri, 1686), about the famous courtesan Yūgiri and her lover Izaemon. Keene has Umegawa sing the piece accompanying herself on the samisen, but in Bunraku today a young girl does the singing.

ji chū u
a. There's no truth in courtesans, people say, but they are deceived,
u
and their words but confessions of ignorance in love. Truth and false-
fushi
hood are essentially one.

ji u
b. Consider the courtesan, so faithful to her lover that she is ready to
u
throw away her life for him—when no word comes from the man
u
and he steadily grows more distant, brood over it as she may, a
u *fushi*
woman of this profession cannot control her fate.

ji chū
c. She may be ransomed instead by a man she does not love, and the
haru
vows she has pledged become falsehoods. But sometimes it happens

that a man favored by a courtesan from the start with merely the
u
false smiles of her trade may when constant meetings have deepened
u
their love, become her lifelong partner; then all her first falsehoods
u *u*
have proved to be true. In short, there is neither falsehood nor truth
iro *u* *fushi*
in love. All we can say for certain is that fate brings people together.

ji
d. The courtesan's love, which seethes and smolders for the man she
suete
cannot meet, may, in the end, burn itself out. Must we curse our

lives!

ji chū u *haru*
e. If we must, then curse it. We can't help loving, that's our sickness.
u
Do all women of our profession have the same chronic complaint?
u
NARRATOR: The story of one who all for love abandoned the world
fushi
induces melancholy reveries, and even the effects of the sake wear off.

Although it appears that the *ji-fushi* pattern in this passage means it is a collection of primary units, because it is one quotation from an earlier Chikamatsu play and is performed as an isolated song, it can be analyzed as a *fushigoto* primary unit divided into five parts. The crux of the first section is that in the world of love, "truth and falsehood are essentially one." Sections b and c discuss how this may be true. The end of section c returns to the conclusion of section a, "there is neither falsehood nor truth in love." In section d, a courtesan's grief and bitterness are expressed in the strongly emotional *suete* musical pattern. In the conclusion, the general plight of courtesans is the subject, and the passage ends with a sad note of submission to a fate too cruel. The unit builds gradually in tension from the general to the particular and from sadness to anger. Finally, it ends on a quiet, melancholic note.

This song, a still moment in a fast-paced drama, introduces the tense, dramatic scene of the "breaking of the money seal," which leads Umegawa and Chūbei to destruction. The line about truth and falsehood being the same is manifested in the following action. Hachiemon enters half drunk and begins to talk disparagingly about Chūbei's dubious business practices and asks the courtesans not to allow him to see Umegawa. Hachiemon sits on the first floor facing the other courtesans, while upstairs behind a door Umegawa listens. Outside the front door Chūbei, too, listens intently to Hachiemon's exhortations. His words in this setting have different meanings for each of the parties who hears them. Chūbei outside takes the words as false slanders on his pride from a friend betraying him; the courtesans, for their part, can only take his words at their face value; it is perhaps only Umegawa who understands Hachiemon's intentions. The scene is masterfully arranged to produce total confusion about Hachiemon's true intentions; the audience, caught up in a whirlwind of contradicting accusations, does not know what to think of Hachiemon and Chūbei's actions.

Learning to feel one's way through a Jōruri play is a new and exciting reading experience. The sense of touch stimulates the mind's eye to probe the life of the text as it was meant to be seen and heard on the stage. The reader caresses the pulse of the rhythm of the actions and sentiments of the characters and is challenged to recreate a full-scale, multisensual drama in the theater of his imagination. Even without having actually heard the sounds of Bunraku, a reader can use the general guidelines of the notation to create his own music to enhance the emotions presented in the text. As a mosaic, the text is static; a reader brings it melody and rhythm and conducts the orchestrated movements of the symphony of emotions. Once brought to life, the language sings in the many voices of human feeling. Music carries these voices—vibrant with the hot blood of passion—into the heart of the reader, as mere cool words without melody could never do.

When one enters into this path [Jōruri]
and tastes its charms, one will never
know the limits of its exotic flavors,
and even the sweetest honeydew will
taste mild in comparison.

Takemoto Gidayū

4 · *Cyclical Imagination*

HISTORY PLAYS

Looking closely at a mosaic is much the same as examining scattered pieces of a jigsaw puzzle; it is impossible to imagine the shape of the whole. We must step back to gain a perspective on its totality—to see how all the parts fit together and form a unified work of art. The mosaic of Jōruri is, of course, not graphic art, but drama with both musical and narrative progression. Each section leads one further along the journey through the crisis of the play until reaching the end. A distinctive characteristic of this journey is that the end is actually a return—after a roundabout route—to the original point of departure.

History plays usually begin in the prominent court or seat of governmental power of the time. In Japanese poetry, from antiquity, the court has been "above the clouds," and therefore Act I takes place metaphorically in a high place. The second act invariably depicts a lower world of conflict and suffering. Not until the fifth act does the play return to the high sphere and to an auspicious conclusion. As in the Western concept of the "wheel of fortune," the play leads to tragedy, the turning point in the drama, although in Jōruri there is a falling and rising pattern rather than the opposite evident in the West. In Jōruri, after tragedy in Act III, the play begins its upward turn, leading to resolution of the problem that fostered the crisis.

Jōruri is not the first genre in which this concept of a cyclical journey appears, in Japan or elsewhere.[1] The prototype in Japan is found in Buddhism, the central tenet of which is the reincarnation cycle of death and rebirth. In this scheme, self-denial, suffering, tragedy, and finally death are steps which can lead one to a new world of enlightenment—free from this recurring cycle. This principle of circularity is fundamental to many Buddhist rituals and iconography, particularly that of the mandala.

An important representation of this circular journey is found in the Main Hall of the Enryakuji Temple (the headquarters of the Tendai sect) on top of Mt. Hiei near Kyoto, where the statue of the Healing Buddha (Yakushi Nyorai) is enshrined. Most statues in Buddhist temples are raised on a platform, and the worshipper looks up to the figure as being in a higher sphere of enlightenment. The icon at Enryakuji is of particular significance because the arrangement of the statue in relation to the viewer is different. The level of the platform upon which the statue rests is no higher than the floor where the follower kneels to pray. This arrangement symbolizes the concept that the heart of everyone is no different and no lower than Buddha: each of us possesses Buddha-nature, though we may not realize it. In this Tendai temple, there is a depression like a deep chasm between the statue and the followers. This ravine must be traversed to reach realization of one's own Buddha-nature. One must journey through various austerities, self-sacrifices, and other experiences to achieve the enlightened stage of Buddhahood. There is a definite sense of returning to the point of departure, because, through the experience of the journey, one realizes that enlightenment was always there; the pilgrim returns to the same point with a new vision of self and universe.

Another structural concept, the Buddhist Six-Realm cycle of existence, is an important organizing principle in Japanese literature, particularly in the popular narrative tradition. The Buddhist idea—of the Six Realms in which the spirit may wander before being freed from the cycle of death and rebirth, and ascending to the upper Four Realms leading finally to Buddhahood—is basic to the popular Japanese imagination. The priest Genshin's (947–1017) treatise, *Ōjōyōshū* (Essentials for salvation),[2] was influential in spreading the tenets of this theory into the mainstream of popular literature. Below the Four Realms of salvation are the Six Realms in vertical ascending order:

Heavenly beings
Mankind
Shura (warrior spirits in battle)
Beasts
Hungry spirits
Hell (with eight levels of suffering)

These realms are within the samsaric world of delusion and desire. Though distinct realms, they also describe states of mind or aspects of personality in individuals.

The idea of "rebirth" after a journey through the Six Realms is clearly seen in the widespread Yamabushi (mountain ascetics) rituals. Carmen Blacker in *The Catalpa Bow,* particularly in two chapters, "The Visionary Journey" and "The Symbolic Journey,"[3] provides a concise outline of the symbolic meaning of the journey of these ascetics to their temples deep in the mountains. Whether the journey was visionary or symbolic, the person returns transformed; that is, reborn with new powers. Her description of the present-day symbolic journey up Mt. Haguro[4] is a fascinating account of a trip through the Six Realms, beginning at the bottom in hell and proceeding upward until reaching the Four Realms of enlightenment. Her summary description of the experience explains the symbolism:

> Thus was enacted, albeit in somewhat dilapidated form, two sets of initiatory symbolism. In a week, we had undergone death, conception, gestation and rebirth in the heart of a mountain which stood for a mother's womb. At the same time we had passed through the realms of hell, hungry ghosts, animals, titans, men, devas, *śrāvankas, pratyeka-buddhas,* bodhisattvas, and finally Buddhas on a mountain which represented a mandala or sacred cosmos. After such an experience the company was in theory endowed with powers beyond the ordinary human allotment.[5]

The iconographical arrangement at Enryakuji represents a journey similar to this Yamabushi ritual, which goes downward through various realms of experience, through symbolic death and rebirth, and finally into Buddhahood. Since in premodern times all classes—from peasants to emperors—consistently participated in temple pilgrimages, often with elaborate prescribed ritual formalities, these circular Buddhist practices with their rather esoteric symbolism did not remain solely the possession of professional monks but found their way into the popular imagination.

An example of the extent of the influence of this symbolism is the *Heike monogatari* (The tale of the Heike), an oral epic which evolved during the thirteenth to fifteenth centuries and was known to all classes throughout the land. Set during the destructive Gempei civil war in the late twelfth century between the Genji (Minamoto) and Heike (Taira) clans, its structure illustrates a Buddhistic journeylike progression of the kind also evident in drama. In simplest terms, the entire tale is a journey to an understanding of the famous opening lines, which state that all is but "a spring's fading dream." The final vision of Kenreimon-in in the "Epilogue," after twelve chapters during which the haughty Heike have been crushed, is both on the same level as the opening lines and is a graphic realization of their significance. Towards the end of the tale when Lady Niidono prepares to dive into the waves at Dan no Ura with the child Emperor Antoku, she says they are going to the "capital beneath the waves." Her words are fulfilled in Kenreimon-in's vision. Her vision of the Buddhist Six Realms is a compressed outline of the entire tale, in which she experiences the highest realm of Heavenly Beings as Mother of the Emperor down to the very depths of hell in battle. This vision is accorded to her only after she too had dived into the sea, been unwillingly saved, and thereby been symbolically reborn. The tale does not end in desolute tragedy, but rather suffering and tragedy lead to a revelation of a vision of unity, tranquillity, and harmony, the new "capital beneath the waves."

The epilogue after the twelve books is called the *kanjō no maki,* which refers to the training ritual a Bodhisattva goes through before becoming a Buddha. This final part of the *Tale of the Heike* is distinct from the rest. As Kenneth Butler has pointed out, many early versions do not have this epilogue; it seems to have evolved gradually during the epic's long development.[6] The tale itself gradually became fixed into a work of twelve books, and the number twelve is obviously important because of the twelve-month year cycle and the twelve-year zodiac cycle in the Sino-Japanese calendar. These uses of twelve delineate complete cycles of time, and the twelve books likewise mark the complete cycle of the rise and fall of the Taira clan. Furthermore, the twelve books symbolize a full cycle in the Buddhist causal chain of existence. In contrast, the epilogue represents a world free from this cycle of destruction and rebirth.

This epilogue is also the subject of the later Nō play *Ōhara gokō* (The imperial visit to Ōhara). The play's climax is the description by Kenreimon-in of her vision of the Six Realms. In most "women" Nō plays the main character (*shite*) is the spirit of a woman of the past who returns to relive the passion of her life through poetry and dance. As in *Nonomiya* (The shrine in the fields) or *Matsukaze,* passionate attachment keeps the spirit from gaining freedom from the reincarnation cycle, and consequent salvation. *The Imperial Visit to Ōhara* contains no struggle or dance, because the character Kenreimon-in (a living figure in the play) has passed the stage of inner conflicts; her experience has given her a vision of the Six Realms which "should not be visible until one has become a Buddha or Bodhisattva," as the former Emperor Go-Shirakawa says.[7] The time is tranquil summer, not the tumultuous autumn of most women plays, and the drama takes place during daytime with no journey into night, the usual pattern. The experience of the "Six Realms," namely the entire *Tale of the Heike,* allows Kenreimon-in to achieve enlightenment.

In the Buddhist popular imagination (with its many shamanistic elements), it is easier to die and be reborn than in the Judaeo-Christian world, and much Japanese popular traditional literature contains stories of reincarnation, symbolic or real. The early collection of Buddhist popular stories, *Nihon ryōiki* (Japanese miraculous stories, 9th century) contains examples of characters traveling to "other worlds" and coming back.[8] Gunji Masakatsu in *Kabuki no bigaku* (Aesthetics of Kabuki) postulates that this concept of transformation or metamorphosis is central to Japanese drama as well.[9] In Nō, the case of a spirit taking a temporary human guise before revealing its true form is a clear example, while in Kabuki the pattern is that a character is disguised and later returns to his or her true self. In the vertical social order of the Tokugawa period (1600–

1868), it was possible through disguise to become free to experience other classes with the idea of eventually returning to one's true station. This movement, however, was only from the top classes downward, and it seems that in Japanese society the top echelons—in medieval times those of the court and in Tokugawa times the high samurai, nobles, and wealthy townsmen—were considered a special class, which in the Buddhist framework of Six Realms is analogous (within the literary world at least) to the Heavenly Being category.

Transformation—or what I describe as a fall from the privileged upper world to the lower realms of trials and tribulations, before the character is "reborn" and returns to his former position—is central to Kabuki. If we use Western terminology, the form could be called "comic" due to its circular structure which ends happily; or perhaps "romance," which contains suffering as part of the comic whole (as in Elizabethan tragicomedy), is a more appropriate term. The Kabuki theater of Chikamatsu's time was based on similar conventions. The *oiesōdō* (disturbance in a grand house) play, as Gunji Masakatsu and others have pointed out, is the most important type of drama in Genroku (1688–1704) Kabuki.[10] The pattern is simple. The heir to a samurai or merchant household is disinherited, due either to his own profligacy in the pleasure quarter or to an evil plotting stepmother or villainous retainer. He goes into hiding, using a disguise (*yatsushi*) as a palanquin bearer, wandering minstrel, or some other humble calling. He bides his time with his lover in the pleasure quarter or is continually harassed, until, in the end, he is restored to his rightful place either by the efforts of a loyal retainer or by his own fortitude. The *yatsushi* portion is the longest, most important part for the actor, during which he displays his histrionic skills. The famous actor Sakata Tōjūrō, for whom Chikamatsu wrote many such plays, was considered a master of *yatsushi.* The pleasure-quarter scenes bring the action to the present, whether or not the setting was in the distant past, as in the case of the *Sukeroku* plays, which are ostensibly based on the twelfth-century story of the Soga brothers' vendetta.

The pattern of *yatsushi* Kabuki—beginning at the socially highest level, moving to a lower sphere such as the pleasure quarter, and returning to the high plane for a happy ending—is

Figure 9. Actor Sakata Tōjūrō (1647–1709) in Yatsushi *Role. From* Ogakuzu *(Sawdust, 1699). Courtesy of Waseda Theater Museum.*

similar to Jōruri. Gunji suggests that Chikamatsu borrowed this *yatsushi* motif from Kabuki for use in his Jōruri history plays.[11] The question of whether Kabuki influenced Jōruri or vice versa is too complex an issue to discuss here. However, if we consider the spatial movements in either theater, we have to search further back to the Nō for antecedents. The *Hachijō kadensho* passage quoted earlier, which outlines the format of a day's program, clearly states that the theatrical experience is a symbolic Buddhistic journey through the torments of hell—which leads ideally to a spiritual awakening. This journey through hell is preceded and followed by higher auspicious settings. The descriptions of Kabuki performances in the *Matsudaira Yamato no kami nikki* (Diary of Matsudaira, governor of Yamato) are similar to Nō and Jōruri in that the theatrical day begins and ends auspiciously. For example, the entry for Kambun 9/8/26 (1669) notes a performance that begins with an Ebisu festive dance of plenitude and ends with an auspicious dance in which all the actors participate.[12] In between these two are short plays (Nō and Kyōgen), dances and farces. Full-length *tsuzuki-kyōgen* (multiple-act plays) gradually become the core of Kabuki programs in the last half of the seventeenth century. The basic plot of these two- or three-act dramas is the *yatsushi* theme, which usually end with grand dances (*sō-odori*) in which the entire cast takes part. They may concern very serious topics of corruption, turmoil, and death, but invariably end happily. This is the convention.

Another structural principle found in Jōruri history plays, which goes along with the convention of cyclical progression, is dialectical opposition between two interacting forces. In simplest terms, the play's basic conflict is between good and evil. In the famous *Battles of Coxinga* (1715), Chikamatsu uses yin-yang concepts of Chinese philosophy to support the dialectic he sets up. Wing-tsit Chan explains the yin-yang philosophy both in terms of a dialectic of forces or principles and of a rotation of the five traditional elements.[13] He explains also that the system is dynamic, and, when applied to philosophy, it envisions political affairs in terms of "a cyclical philosophy of history." These two concepts of the dialectic and the cyclical are fundamentals of the human imagination; language is basically structured on a principle of opposition (dark-light, good-bad), and our routines of life revolve on the daily, monthly, and yearly cycles of nature, principles familiar to the West. Northrop Frye has formulated a theory of literature based to a large extent on Western conceptions of imagination not dissimilar to those discussed above.[14]

Chikamatsu, when writing Jōruri history plays, necessarily worked within conventions he inherited from the tradition. Musical conventions determine the narrative structure and mode of rhetoric and complement the conventions of dialectics and cyclical movement which shape the play as a circle in which tragedy is the turning point towards a happy conclusion.

THE SOGA HEIR

A detailed examination of four of Chikamatsu's five-act history plays demonstrates how theory and practice converge. These examples have been chosen both for their relative importance in Chikamatsu's development as a playwright, and because they were written for various chanters. The first two plays, *Yotsugi Soga* (The Soga heir, 1683) written for Uji Kaganojō, and *Shusse Kagekiyo* (Kagekiyo victorious, 1685) written for Takemoto Gidayū, are the earliest works known for certain to be by Chikamatsu. We shall examine them primarily in relation to the writings of the two chanters to show how Chikamatsu molded his narrative to fit their ideas. Later plays, *Semimaru* (1688), also written for Gidayū, and particularly *Kokusen'ya kassen* (The battles of Coxinga, 1715), written for Takemoto Masatayū (Harima no shōjō) and Gidayū's other successors, typify mature plays by Chikamatsu, who gradually gained control of the conventional form and used it to create increasingly successful plays.

In editions of *The Soga Heir* published during Chikamatsu's lifetime only Kaganōjo's name appears, but other evidence proves that it is by Chikamatsu. We will view this drama from Kaganojō's perspective, because Chikamatsu had to write to satisfy his employer. The latter two plays have been translated into English, but unfortunately the former two have not, though there is a German translation.[15]

The five-act *Soga Heir,* set in the twelfth century, begins with the successful attack of the Soga brothers to avenge their father's death, after having patiently waited eighteen years for the chance. The older brother Jūrō is killed, and Gorō is taken captive and later executed. The second scene is a musical (*fushigoto*) section in which the spoils of the hunt are catalogued, and the Soga brothers too are listed among the fallen beasts. In the final dramatic scene, the mighty Asahina strongly defends the dignity of the Soga brothers against the villains who listed the brothers as beasts. But Yoritomo calms Asahina and preserves the festive atmosphere.

In Act II, the loyal Soga retainers, the brothers Oniō and Dōsaburō, are introduced. They are sincere but bumbling clown-heroes. Upon hearing the Soga name slandered, both vow revenge but argue over who is to go, and finally they part in

Figure 10. Scene from The Soga Heir *(Yotsugi Soga), Act I, Scene i. Lower right shows Mt. Fuji and the Soga brothers after Jūrō's death. Upper left shows Yoritomo deciding Gorō's fate. Courtesy of Waseda University Library. Photograph courtesy of Shōgakkan.*

anger. In the lyrical second scene, we meet the Soga brothers' courtesan lovers, Tora and Shōshō, who are anxiously awaiting news of the vendetta. In the next scene the villains, who had slandered the brothers, try to seduce the girls but are driven out by Oniō. In the final scene, Oniō tells of the deaths of the brothers. The girls attempt suicide but are persuaded to live on and to pray for the salvation of their lovers. Oniō then sets out after the villains.

Act III has a more serious atmosphere comparable to Act I. The girls make a long tearful journey (*michiyuki*) to the home of the aged Soga mother. They visit the sick woman and, out of sympathy, pretend to be her sons. In the final scene, they relate the revenge attack and the mother realizes her sons are dead. She is, however, consoled by the revelation that Gorō has a son.

Act IV is filled with excitement and humor in contrast to solemn Act III. The two Soga retainers again quarrel over who is to give chase. The next scene shifts to the villains, who come to

Tora and Shōshō, demanding the heir, but are tricked into hiding in chests in order to attack Oniō and Dōsaburō. However, Asahina arrives first and hoists two boulders onto the chests, crushing one villain, and captures the other. Act V returns to Yoritomo's presence where the villain is pardoned and made to become a priest to pray for the Soga brothers. The play concludes with an auspicious dance of the courtesans. All ends in harmony.

An analysis based on the musical notation allows us to see the play as Kaganojō (and the playwright Chikamatsu) envisioned it. The five acts are organized on a principle of inner balance and tension: Acts I and V are complementary, as are II and IV, while III stands alone in the middle. The main action of the first act is Gorō's supreme bravery in the face of death; in the final act, Tora (Gorō's wife) resolves the conflicts of the entire play by showing mercy in pleading for the pardon of the surviving villain. The story comes full circle: from Yoritomo's presence, down through conflicts and action, and back to Yoritomo at the end. The cycle is complete after the transition in Act V from revenge to mercy. The dialectical forces revolve; each has its moment of importance before giving way to the other.

Using the notation as a guide, we can break the play down into its components to see how the narrative is organized. Act I can be analyzed as follows. (The page: line references are to the *Chikamatsu zenshū*, vol. II, edited by Fujii Otoo).

Three scenes: i 8 primary units; two parts
a. p.1: line 2 to *okuri* 6:2
b. 6:2 to *sanjū* 10:4
ii 5 primary units 10:4 to *sanjū* 14:6
iii 2 primary units 14:6 to 16:8

Scene i begins with a description of the glory of the Kamakura shogunate under the leadership of Yoritomo, but this lofty atmosphere quickly changes as the scene shifts to the revenge attack, the death of Jūrō, and the capture of Gorō (to the *okuri*). The first two units are introductory to the remaining six of this first scene. The story told is of the courageous Gorō, defiant even towards the Shogun Yoritomo; and Gorō's brave death is the climax. The final short unit of this scene is transitional, and is festive and nondramatic. Scene ii has three movements divided by the *okuri* notation. The passage is recited by two chanters, and since it contains no *kotoba,* it is a (*fushigoto*) song in contrast to the actively dramatic *jigoto* Scene i. The last unit, in which the Soga brothers are referred to as beasts, leads to Scene iii in which Asahina, on behalf of the insulted Soga brothers, threatens revenge against the villains, introducing the main conflict of the drama.

The middle festive section might seem out of place to modern readers if they were not aware of Kaganojō's principle of musical variety. In this context, the light-hearted Scene ii does make sense. At the beginning of the act, the audience sees a quick battle, many deaths, and the execution of a proud Gorō. The festive "catalogue of hunting spoils" song relaxes the tension. The short final scene, however, rejuvenates the suspense, making the audience anxious to see the outcome of Asahina's threat.

The musical styles of the primary units contain a variety of both musical tone and dramatic content:

i a. introductory units 1 and 2 are in *jigoto* style without *kotoba*
b. units 3–7 are basically dialogue (mostly in *kotoba*)
ii units 8–13 are sung (*fushigoto*)
iii units 14 and 15 are mainly in *kotoba*

Although in terms of content (a catalogue of hunting spoils), the middle scene is hardly interesting for a reader, it was the point when the chanter would display his talents, and, therefore, was enjoyable for the listener. In later *danmonoshū,* collections of scenes, this section together with the third-act *michiyuki*, the third-act reenactment of the Soga revenge by the two women, and the fifth-act final dance, are the scenes included from the play, rather than the more "dramatic" ones. The *danmonoshū* of Gidayū also contain the musical scenes from this play, not the dramatic ones. We can get an idea of the later change of preference by amateur chanters, if we notice sources and choices of practice texts in other periods when amateur chanting was popular during the late eighteenth century and again in the first quarter of the twentieth century.[16] Popularity shifted to the more dramatic tragic scenes. The connoisseur audience of the early period (1670s to 1720s), at least as their tastes are reflected in the *danmonoshū,* consistently preferred musical over dramatic scenes. Both musical-lyrical and dramatic aspects of the plays were entertaining for them.

Unlike Act I, the atmosphere of Act II is not noble, but rowdy and comic, and not in Yoritomo's presence, but mostly in the pleasure quarter. This discontinuity might be disconcerting if we were unaware of Kaganojō's precept of variety of styles. Act II is presented to contrast with Act I as well as with Act III. The act is not in just one setting but moves to a different location at each scene change. Villains, courtesans, and bumbling retainers also contrast as a cast of characters to the courtly hunting party. Even within Act II there is variation of mood. Scene ii is lyrical and was sung primarily in the *ji* melodious style with no *kotoba* at all. The subject is the pitiful love of the courtesans for the Soga brothers. This sad theme of their futile love becomes central in Act III. The atmosphere of the *michiyuki* and the whole third act is sad and dark, lyrical rather than dramatic. Act IV returns to dramatic action and essentially takes up the story where Act II left off. Act V returns to the noble level of Act I. Tora pleads with Yoritomo to pardon the surviving villain and have him become a priest to pray for the Soga brothers. Finally, the atmosphere becomes festive when Yoritomo's wife asks the girls to dance.

In contrast to the other three acts of the play, Acts II and IV are light and active. In most five-act history plays, these two acts contain battle or fight scenes. Such contrasts within one work fit Kaganojō's aesthetic scheme, because his principle is that each part must have a distinctive musical style. Therefore, by definition Acts II and IV must contrast with the acts around them. In *The Soga Heir,* the dramatic conflicts introduced in Act II are then resolved in Act IV. The characters Oniō and Dōsaburō, the loyal Soga retainers, likewise appear only in these two acts. Acts II and IV alone concern the dramatic theme of the vindication of the Soga honor. The rest confront the universal theme of the revenge cycle and the release from it through sacrifice and mercy.

This framework of balance centers on Act III, which stands alone. Most often Act III contains tragic death. In *The Soga*

CHART 2
Musical Structure of *The Soga Heir,* Acts I and II[a]

ACT I, 3 scenes

i Opening to *sanjū* 10:3
 1. a. Glory of Minamoto power (to *oroshi* 5:7)
  b. Yoritomo's hunting camp (beginning 5:7)
 2. Revenge of Soga brothers: Jūrō is killed, Gorō captured (5:10 to *okuri* 6:1)
 3. Yoritomo and Gorō dialogue; valiant Gorō (6:2)
 4. a. Gorō explains his reasons for attacking Yoritomo's men (6:12 to *suete* 7:12)
  b. Valiant and defiant Gorō (7:12)
 5. Yoritomo is impressed with his valor (8:2)
 6. a. Jūrō's head is brought; Gorō weeps (8:7 to *suete* 8:10)
  b. Gorō's love for his brother (8:10)
 7. Yoritomo wishes to pardon him, but Gorō must die; yet his glory will live on forever (9:4 to *okuri* 9:12)
 8. Transitional passage (9:12 to *sanjū* 10:3)

ii Two chanters: musical (festive) section (10:4 to *sanjū* 14:6) (*fushi-goto*). List of the spoils of the hunt. Five units; in the last one the Soga brothers are listed as beasts among the spoils of the hunt (13:9)

iii Final dramatic scene (14:6 to end of the act)
 14. Asahina stands up for Soga dignity and threatens villain, Arashirō (14:6)
 15. Yoritomo stays Asahina saying that the time is auspicious and festive. Asahina ends the act with a threat to Arashirō (15:10)

ACT II, 4 scenes

i 16:11 to *sanjū* 20:3
 1. Introduction of loyal Soga retainers, the brothers Oniō and Dōsaburō (16:11)
 2. Asahina reports news of slander
  Brothers argue about who goes for revenge (comic) (17:4)
 3. Brothers part in anger

ii 20:3 to *okuri* 22.6 (contains no *kotoba*)
 4. Lyrical description of plight of courtesan lovers of Soga brothers, Tora and Shōshō (20:3)
 5. Continuation (20:8)
 6. No word from Soga brothers (20:11)
 7. Love of courtesan (Tora) for Gorō (21:8)

iii 22:6 to *okuri* 25:4
 8. Villains are introduced; they worry about Asahina's threat (22:6)
 9. They try to seduce Tora and Shōshō (22:9)
 10. Oniō drives the villains out (23:6)

iv 25:4 to the end of the act
 11. Oniō explains success and death of Soga brothers (25:4)
 12. Lyrical: girls attempt suicide; instead decide to pray for the souls of the Soga brothers (25:5)
 13. Oniō sets out after the villains

Note: [a]Small roman numerals mark scene divisions; arabic numerals, primary units. See glossaries for explanation of terms. Page references are to Fujii, *Chikamatsu zenshū,* vol. II.

CHART 3
Outline of the Five Acts of *The Soga Heir*[a]

ACT I Setting: Shogun Yoritomo's camp, noble, little action, all male characters

i	ii	iii
jigoto	*fushigoto*	*jigoto*
High noble atmosphere; Gorō's virtue	Festive; musical	Dramatic; threatening

ACT II Setting: pleasure quarter, lower, comic, active and heroic, men and women

i	ii	iii	iv
jigoto	*fushigoto-like*	*jigoto*	*jigoto*
Comic farce	Lyrical	Crass villains	Heroic

ACT III Setting: Journey, lyrical, sad, all women, no *kotoba* in whole act

i	ii	iii
fushigoto (michiyuki)	*jigoto*	*jigoto and fushigoto*-like dance
Tortuous journey; downward movement and dark images	Sick Soga mother	Story of Soga revenge, victory, death, hope for child heir

ACT IV Setting: Common world, dramatic action, men and women

i	ii	iii	iv
jigoto	*jigoto*	*jigoto*	*jigoto*
Feuding brothers	Chase villains; emotional reunification, turning point in play	Villains plotting	Dramatic victory over villains

ACT V Setting: Court, noble and auspicious, Yoritomo and Tora dialogue

i	ii
	furyū dance of unity harmony
jigoto	*fushigoto*
Surviving villain is pardoned; festive scene; Tora (Gorō's lover)'s virtue	Love of courtesan; mother and child; final vision of plenitude and harmony

Note: [a]Small roman numerals mark scenes.

Heir, the deaths are symbolic; the two women experience death through the medium of dance in the reenactment of the Soga vendetta. Gidayū refers to the pathos of Act III as "the essence of the entire play," and Chikaishi Yasuaki has in more recent times developed a thesis about Jōruri, using as his central phrase, "Jōruri, an art of pathos" (*urei no geijutsu*).[17] Act III is usually where the central pathos is found. After the suffering in Act III, however, the play begins to ascend towards the happy ending.

The structure of balance and contrast is also reflected in other ways. In the Soga play the opposition of revenge and mercy is supported by an opposition of male and female. Chart 3 lists whether women or men, or both, dominate the action of each act. The first act contains only male characters, the third act only women, and the final act is dominated by merciful Tora and completed by the dance of harmony performed by the two courtesans. Usually in the third act a change occurs in the play's direction; in this play the act is lyrical and filled with pathos. It begins with Tora and Shōshō's journey, during which they lament the loss of their lovers. The opening lines of the *michiyuki* are:

> Love drives all to madness
> And to depths of pain and confusion.
> We are fallen women from the mountains,
> Tossed about in a floating world of dreams,
> Enticed by melodies of pleasure.

The movement is imagistically downward into a dark world of passion and suffering. Further along the journey there are references to the "burning house" and other Buddhist images of suffering. The two women plead that their suffering is worse than Buddha's when he abandoned his family, because their lovers are dead. The second scene takes place in the Soga home, where the ailing mother is attended by her sister. The girls decide to disguise themselves as the Soga brothers. The time is evening and, since the mother is almost blind, she embraces an illusory hope that her sons are alive. The final scene is the "Jūbankiri" (Revenge attack of the Soga brothers). It is a dance version of the attack and in it the chanter Kaganojō, and later Gidayū, who also performed the play, were given ample opportunity to display their skill in an exciting battle pageant. No one actually dies in this third act, which is unusual for a five-act history play.

If one has read no other Jōruri play than *The Soga Heir,* then it might be hard to accept an interpretation of this dance version of the vendetta as being a symbolic experience of death for Tora and Shōshō. It may appear that the interpretation of viewing the third-act experience of Tora and Shōshō as symbolic death is an over-reading of a much simpler play. *The Soga Heir,* however, does contain the outline of downward and upward movement (and of death and rebirth) that Chikamatsu clearly makes fundamental in his later works. If this play were judged solely by Aristotelian or Elizabethan standards of dramatic unity or character development, then it might hardly appear comprehensible as a work of art. However, within the conventions of its own tradition, the play, while certainly not a masterpiece, is nevertheless an intricately structured piece founded on a principle of musical and dramatic orchestration.

Chikamatsu had to produce coherent plays to fit the chanters' musical formulas. The chanter saw each part in both musical and dramatic terms, and the playwright had to match the imagery to the musical and dramatic mood. One way Chikamatsu creates a poetic mood for each act is by giving them different "spatial settings." The first act, the highest plane in the play, is set in the court of the period or the seat of the governmental power. The Soga play begins in Yoritomo's presence, a suitable setting because he is the ruler of the land. The world of Acts II and IV is the pleasure quarter, a place at the bottom of the social spectrum. Act III is a world lower than any of the other acts, and can be likened to a hell of suffering. The movement between these five acts is cyclical; there is a downward movement to the lowest point in Act III, which is then followed by an upward movement to the same setting as the beginning.

The Soga Heir fits neatly within the theory of Jōruri outlined by Kaganojō in the *Collection of Bamboo Shoots.* The convention is a pattern of musical orchestration with pathos as the central focus. The audience is gradually led to the crisis by wavelike movements, but the drama does not end in disaster. The finale is always an upward movement to a world of joy and harmony.

This very early Chikamatsu work also introduces the elements of dialectics and cyclical movement, which were later used by the playwright with greater skill as he came to master the conventions to create more interesting dramas.

KAGEKIYO VICTORIOUS

Shusse Kagekiyo (Kagekiyo victorious, 1685) is the first play we know for certain Chikamatsu wrote for Takemoto Gidayū. The time setting is the late twelfth century, generally the same as for the Soga play. Kagekiyo is one of the few survivors of the Taira clan, who were defeated by the Minamoto under the leadership of Yoritomo. In this play Kagekiyo is trying desperately to avenge the Taira defeat by assassinating Yoritomo. The opening is an evocation of the mercy and power of the Bodhisattva Kannon of whom Kagekiyo is a devout worshipper. Kagekiyo is in hiding and marries Lady Ono, the daughter of a Taira sympathizer who has helped him. The second scene is a song, a "catalogue of materials"; this time the materials are for the construction of the great Nara temple Tōdaiji, which is being rebuilt by Yoritomo. It is the ground-breaking ceremony, and therefore auspicious. In the final scene, Kagekiyo, disguised as a worker in order to attack Yoritomo, is discovered by Yoritomo's general and forced to flee, after fighting and killing fourteen or fifteen of the enemy. The scene must have been visually exciting, because Kagekiyo throws all the carpenter's tools into the air, announces a threat to return and exact revenge, and then flies away!

Act II begins with another reference to Kagekiyo's devotion to Kannon. This act introduces Kagekiyo's courtesan-wife Akoya, a passionate woman who becomes violently jealous when she learns of Kagekiyo's new wife. To win a reward, her brother exploits this jealousy to get her to reveal Kagekiyo's location. She betrays Kagekiyo but immediately regrets her action. The final scene is an attack of 600 soldiers against Kagekiyo at a temple; he fights them off and again escapes by flying away.

In Act III, Lady Ono's father Daiguji is captured, and Lady Ono makes a journey (*michiyuki*) to find Kagekiyo but is also captured. In the final scene we see her steadfastness in the face of torture. She refuses to betray Kagekiyo; he surrenders to save her life.

Act IV finds Kagekiyo behind the walls of a mighty prison, praying fervently to Kannon. Lady Ono's virtues are extolled. In the second scene, Akoya returns and pleads to Kagekiyo for forgiveness. Kagekiyo, however, refuses to acknowledge her presence. In atonement for her action she kills their two sons and finally herself in an intensely dramatic scene. After this, the mighty Kagekiyo breaks loose from the prison, but returns and surrenders again to spare Lady Ono and her father. In the final glimpse of Kagekiyo, we see him praying to Kannon.

In Act V, we learn that Yoritomo has given orders for Kagekiyo's execution, but an argument ensues over whether he is in fact dead or not. An investigation reveals that he is indeed alive and that, mysteriously, the head of a Kannon statue has been severed and blood spilled about. This miracle impresses Yoritomo, who praises Kannon and pardons Kagekiyo. The final scene is Kagekiyo's dance before Yoritomo in which he reenacts his battle adventures. During the dance he raises his sword, as if to strike Yoritomo, but instead tosses it aside. To keep from ever seeing Yoritomo again and having his ire aroused, Kagekiyo plucks out his eyes and gives them to Yoritomo. The last words of the play are for eternal peace in the world.

This play has been praised by a postwar Japanese critic, Hirosue Tamotsu, for its "tragic" proportions and as the first "tragedy" of the Tokugawa period. Actually, although the play does contain tragic elements, its structure is quite similar to *The Soga Heir,* which an analysis of the play into its components demonstrates. A comparison of the first acts of *Kagekiyo Victorious* and *The Soga Heir* shows that similar narrative patterns were used. The opening preludes are lofty and suggest the outline of the first and fifth acts. The opening of *Kagekiyo Victorious* is an invocation to Kannon Bodhisattva's power and mercy; *The Soga Heir,* on the other hand, begins with praises for the reign of the Genji under the leadership of Yoritomo. Kagekiyo's devotion to Kannon is one of the unifying themes of the whole play, because it is from Kannon that Kagekiyo obtains his supernatural powers, and later his life. Yoritomo's august presence and thoughtful judgments both begin and end the Soga play.

The first acts of both dramas divide into three scenes according to the notation, and both middle scenes are festive songs.

CHART 4

Musical Structure of *Kagekiyo Victorious,* Act I[a]

ACT I, 3 scenes

i Opening to *sanjū* 27:13
  1. a. Prelude: mercy and power of Kannon (to *oroshi*)
     b. Valiant Kagekiyo is in hiding, waiting for a chance for revenge against Yoritomo (beginning 26:7)
  2. a. Daiguji helps him; Kagekiyo marries his daughter, Lady Ono (26:11)

ii 27:13 to *sanjū* 29:12 Musical section (*fushigoto*)
  3. Preparation ceremony for rebuilding Tōdaiji, auspicious mood (27:13)
  4. Catalogue of materials; praise grand building to be created (28:9)
  5. Final flourish; jewels, gold, and silver cover the building; toast to commencement of construction (29:4)

iii 29:12 to end of act
  6. Kagekiyo lurks about disguised as a worker (29:12)
  7. Shigetada, Yoritomo's general, recognizes Kagekiyo (30:8)
  8. Kagekiyo slays 14 or 15 men, throws all the tools in the air (31:3)
  9. "I will return to seek revenge," and he flies away (31:8)

Note: [a]Small roman numerals mark scenes; arabic numerals, primary units; and small letters, subdivisions of primary units. See glossaries for explanation of terms. Page and line references are to *Chikamatsu jōruri shū,* II, edited by Shuzui Kenji and Ōkubo Tadakuni.

The songs, however, have different roles in each case. In Kaganojō's play, the song takes place towards the end of the act and therefore is most important, but in Gidayū's play the song is prologue to the dramatic action at the end, when Kagekiyo displays his prowess. This difference suggests that Chikamatsu was influenced by the more dramatic interests of Gidayū rather than the musical emphasis of Kaganojō. Gidayū notes in his treatise that, in the first act, "one should chant vigorously as if manipulating loose and wild reins, and strive to entice and excite the feelings of the audience." He also emphasizes that the end of the first act is important in determining the success or failure of the performance. The final battle scenes of the first act of *Kagekiyo Victorious* are certainly calculated to excite. The *sanjū* notation normally signals the conclusion of a scene, but it also is used during battle scenes. In the final flight scenes, Gidayū uses two *sanjū* sequences to build up the action to successive crescendos. The final flourish concludes on an exciting note,

CHART 5
Outline of Five Acts of *Kagekiyo Victorious*[a]

ACT I Setting: Temple, noble mood, devotion of Kannon, prowess and determination of Kagekiyo for revenge

i	ii	iii
jigoto	*fushigoto*	*jigoto*
Plight of Kagekiyo; marries Lady Ono; determination for revenge	Auspicious; festive	Dramatic revenge attempt; prowess of Kagekiyo; flies away

ACT II Setting: Commoners' world, courtesan Akoya's betrayal, Shura action

i	ii	iii	iv
jigoto	*jigoto*	*jigoto*	*jigoto*
Devotion to Kannon; Akoya's love	Akoya's jealousy	Akoya's betrayal due to jealousy	Kagekiyo's power; 600 enemies; Kagekiyo flies away

ACT III Setting: Lady Ono's *michiyuki,* torture, lyrical, sad

i	ii	iii
jigoto	*fushigoto*	*jigoto*
Daiguji (father-in-law) captured	Lady Ono's journey and capture	Devotion to Kannon; Lady Ono is firm under torture; Kagekiyo surrenders

ACT IV Setting: Kagekiyo in prison, Akoya's tragedy, dramatic

i	ii	iii
jigoto	*jigoto*	*jigoto*
Mighty prison; pray to Kannon; pure Lady Ono	Intensely dramatic; Akoya kills 2 sons and herself; powerful (violent) emotions of Akoya	Kagekiyo breaks loose but returns to save wife and father-in-law; prays to Kannon

ACT V Setting: Yoritomo's court, Kannon's mercy and power, Yoritomo's mercy, Kagekiyo's passivity

i	ii	iii	iv	v
jigoto	*jigoto*	*jigoto*	*jigoto*	*jigoto and final fushigoto*
Yoritomo	Dialogue; Kagekiyo alive or dead?	Investigation; power and mercy of Kannon	Yoritomo's pardon for Kagekiyo	Kagekiyo's battle tale (Nō); gives up sword (*Atsumori* pattern); praise to Kannon

Note: [a]Small roman numerals mark scenes.

as the superhuman Kagekiyo battles the swarming enemy and finally, when trapped, flies away into the heavens.

The overall organization of *Kagekiyo Victorious* follows a conventional pattern similar to that of *The Soga Heir.* Although the plot revolves on the struggle of Kagekiyo in seeking revenge against Yoritomo, the story divides into two subplots around the women, Lady Ono and Akoya. Acts I, III, and V are closely related, and Acts II and IV are likewise intertwined. Dramatic tension is maintained throughout. Acts I and II end with Kagekiyo fighting for his life and finally being forced to flee, literally flying away; Acts III and IV end with his surrender and fate in doubt. This strategy follows Gidayū's emphasis in his treatise on the importance of the endings of the acts. The conflicts are not completely resolved until Act V, when Kagekiyo—in Nō dance—relives the destruction of the Heike, gives up his sword, blinds himself, and purges his desire for revenge. This technique of maintaining dramatic tension throughout the five acts is altogether different from *The Soga Heir,* in which each act ends with hope.

The most influential interpretation of *Kagekiyo Victorious* in Chikamatsu criticism is *Chikamatsu josetsu* (Introduction to Chikamatsu) by Hirosue Tamotsu.[18] His theme, the development of tragedy in Chikamatsu's plays, has stimulated Chikamatsu scholarship since its inception in 1957. It is perhaps the only critical work on Chikamatsu that has gone through several editions. While Hirosue's work is of note because of its impact on the field, his critical methods, which do not take into account Jōruri traditions, are often misleading. Hirosue postulates a priori that there must be a natural development toward "tragedy." Chikamatsu's plays are shaped to fit into this hypothesis. Tragedy is considered the ultimate goal of theatrical development. Hirosue's central concept is that the action of the main characters moves to an impasse caused by a sole or comprehensive complication which controls the fates of the characters. Death alone can break this impasse. His central phrase, "*kattō no tan'ichi na renkan,*" means tragedy "is directed by a single complication."[19] Upon this Aristotelian-like theory he bases his thesis that Chikamatsu's *Kagekiyo Victorious* is unique in Japanese dramatic and literary history, and an epochal step out of the world of "old" Jōruri (*ko-jōruri*) into a new tradition of dramatic tragedy.

Some fundamental problems with Hirosue's methods give birth to misunderstandings of Chikamatsu's plays. Initially, his approach suggests that his concept of "tragedy" is the basis of Chikamatsu's art, but Hirosue's *higeki* (tragedy) is a nineteenth-century translation of the Western word for dramatic tragedy, which expresses a concept unknown to Chikamatsu. Although Chikamatsu's idea of pathos as restraint on the characters is similar to Hirosue's Western-oriented theory, there are significant differences. As we have seen, pathos is only part of the total experience of a day of theater, not the end in itself. When Chikamatsu's plays fail to conform to Hirosue's definition of tragedy, he takes the playwright to task for immaturity. If the plays do not fit the theory, which is alien to Chikamatsu's world, then the approach should be questioned, not the play. Hirosue's focus, solely upon those tragic elements which fall within the sphere of his theory, prevents him from criticizing the plays on their own terms.

Hirosue suggests that *Kagekiyo Victorious* has two major themes which are not unified into a coherent whole.[20] One is the complications between Akoya and Kagekiyo; the other is Kagekiyo's struggle for revenge against Yoritomo. He focuses on the Akoya tragedy as the crux of the play. It is certainly true that Akoya's tragedy of killing her (and Kagekiyo's) two children and then herself as retribution for betraying Kagekiyo is exciting drama, and something new in Jōruri. Akoya, through her violent passions of love, jealousy, and finally pride, comes to life as a full-bodied character. However, the Akoya story is only a part of the drama. Hirosue's analysis moves along without encountering too many obstacles until the final act in which there is a miracle as well as a happy ending. He considers such miracles a hindrance in the development toward realistic tragedy.[21] This dogmatic attitude makes it impossible for him to make sense of the final act or of the role of Lady Ono, whom he virtually ignores, although her sad *michiyuki* journey, capture, and torture occupy the important third act. The final references to the Nō *Kagekiyo* are considered by Hirosue to be merely tacked on and to have no connection with the rest of the play.[22] In short, according to his analysis, the tragedy of Akoya is fascinating, but the play as a whole is riddled with structural faults. He misses the pattern of the mosaic by focusing too narrowly on a few parts.

However, as we have seen, *Kagekiyo Victorious* is also organized on a principle of musical variety. Akoya's tragedy, Kagekiyo's struggle, and Lady Ono's patient suffering, while being isolated incidents in the play, are also unified by a larger theme holding the play together—from beginning to end. Since Kagekiyo is important throughout the play, let us examine the background of this larger-than-life Heike warrior. Two strands of his legend describe different, even contradictory, aspects of Kagekiyo's character. One legend (the subject of the Nō play *Daibutsu kuyō* [Dedication ceremony at the Great Buddha Hall], the Kōwaka-mai dance play *Kagekiyo,* and the old Jōruri play *Kagekiyo*) holds that, as a survivor of the Gempei wars, Kagekiyo tried many times—thirty-seven in all—to avenge the Heike by killing Yoritomo. The other strand of this legend (the subject of the Nō play *Kagekiyo*) is the story that Kagekiyo became a blind beggar who turned his back to the world and to his own past. The main source for *Kagekiyo Victorious* is the heroic, revengeful aspect of the legend found in the old Jōruri *Kagekiyo* and the Kōwaka-mai *Kagekiyo,*[23] while the ending incorporates the scene from the Nō *Kagekiyo* in which he retells the tale of his battle glory. His memories are of victorious battles and his final recollection is of his fight with the Genji warrior Mionoya Jūrō at Yashima, when Kagekiyo is fearless. This is the last victorious battle of the Heike before Yoshitsune defeats them at Yashima.

These two aspects of the Kagekiyo legend stand in opposition. One depicts an unrelenting, revengeful, and passionate Kagekiyo; the other portrays a blind, religious, and submissive Kagekiyo, who seeks freedom from his attachments and violent passions. Chikamatsu portrays the two female characters in Kagekiyo's life, Akoya and Lady Ono, as contrasting personalities, paralleling these opposing aspects. The figure Akoya (Akoo) had appeared in the earlier Kōwaka-mai play as a woman who betrayed Kagekiyo and was killed as punishment for her actions. In the old Jōruri play she is specifically a courtesan. Chikamatsu makes her a violently passionate character who betrays Kagekiyo due to jealousy. (In the old Jōruri she had done it with hopes of gaining favor for the children, but Kagekiyo killed the children.) She immediately regrets this and eventually destroys both her children and herself before Kagekiyo's eyes, as a plea for forgiveness. The other woman, Lady

Figure 11. Ukiyoe Actor Print (Kuniyoshi, 1849) of Kagekiyo (Ichikawa Danjūrō). Courtesy of Waseda Theater Museum.

Ono, Kagekiyo's official wife, had appeared as nothing more than a name in earlier plays, but Chikamatsu makes her a figure of resolute purity, fortitude, and patience in contrast to the violent, impetuous Akoya. Although Hirosue ignores Lady Ono's role, her dominant position in Act III alone gives her importance, and the torture scene, due to its location at the end of Act III, is filled with pathos, which Gidayū a few years later described as the essence of the entire play. Shinoda Jin'ichi argues that, since it is mainly Lady Ono's *michiyuki* scene that appears in the *danmonoshū* collection of scenes, the audience of that time considered her role important.[24] Chikamatsu has the two women characters stand in opposition. Akoya is comparable to the legendary Kagekiyo, who brashly and violently sought revenge for the Heike defeat, while in contrast Lady Ono's character is analogous to the enduring passive Kagekiyo, the blind beggar, who bears the weight of the dark world alone in his hut.

Hirosue found the miracle—of Kagekiyo's death and resurrection through Kannon's mercy and power—a flaw in the play and considered it a remnant of the medieval world that Chikamatsu should have discarded.[25] This kind of attitude has been expressed about Shakespeare's fanciful romance plays as well. Ben Jonson was perhaps the first to refer to them as "mouldy tales." Rather than simply to consider use of the supernatural as regressive, Chikamatsu's use of the miracle play should be examined within the context of the theatrical conventions.

Shinoda suggests that Chikamatsu's opening of *Kagekiyo Victorious* with an invocation to Kannon puts the form in the *sekkyō* Buddhist miracle-play tradition in which a Bodhisattva dies to save the main character.[26] An early example of this type is *Amida no munewari* (Amida's riven breast, 1630). In *sekkyō* Jōruri there was also a play with the title *Kagekiyo,* but the text has not survived. Chikamatsu has chosen this archetype found in the Buddhist miracle play as the context for his dramatization of the legend.

The theme of dying for another is common in Jōruri. Kuroki Kanzō, in a discussion of "Terakoya" (The village school) as a scene based on this convention, describes three kinds of *migawari,* "sacrifice for another."[27] The first he puts in the religious context, that of a Bodhisattva dying in place of someone who has shown extreme devotion, piety, or purity of heart. He suggests *Amida's Riven Breast* and *Kagekiyo Victorious,* among others, as examples. His second category is sacrifice for love; usually a woman sacrifices herself to help her husband or lover. It is interesting that he puts Akoya's suicide in this category.

Figure 12. Scene from Amida's Riven Breast *(Amida no munewari, 1680s).* Eiri-jōruri *picture book. Through Amida's compassion, the girl's life is spared. Blood gushes forth from a statue of Amida. Courtesy of Waseda Theater Museum.*

The third he classifies as sacrifice for ethical reasons. The usual pattern is the suicide of a retainer for his lord.

Amida's Riven Breast is representative of the Bodhisattva-sacrifice type.[28] Two children are left abandoned when their rich parents are executed for crimes. They decide to sell themselves in order to raise money for offerings for their parents' souls. There is a rich man whose son is sick and can only be cured by administering the liver from a maiden born at the same time as the son. The wandering children go to this man, and the young girl offers herself in exchange for a promise that the rich man will build a temple to Amida for the salvation of her parents. Amida has mercy; the next day the girl is safe, and the Amida statue splits open and blood gushes forth.

In *Kagekiyo Victorious* Chikamatsu uses this Bodhisattva-sacrifice convention to incorporate the two contrasting Kagekiyo legends in one play. Only after the passionate, violent Kagekiyo dies is the passive, merciful, and patient Kagekiyo reborn. As in *The Soga Heir,* the overall structure is founded on dialectics and cyclical movement: revenge versus mercy, jealous Akoya versus patient Lady Ono, and passionate, revengeful, warrior-like Kagekiyo versus merciful, submissive, and self-blinded Kagekiyo. After the self-sacrifice of the jealous Akoya, Kagekiyo relinquishes his desire for revenge and instead sacrifices himself to save the passive Lady Ono—a decision which carries him to the world of mercy in Act V, where he is freed from the recurring cycle of revenge. In Jōruri, tragedy is both an end and, at the same time, a beginning.

SEMIMARU

In the later *Semimaru,* Chikamatsu makes clearer what this new beginning can mean. This play, first performed in 1688, was revived at least two times during Chikamatsu's lifetime, one of which was on the occasion of Gidayū's receiving an honorary title from the court. Susan Matisoff, who has translated this work, explains that Chikamatsu made use of the various aspects of the Semimaru legend while adding elements to the basic story, a practice which Zeami had followed in writing Nō plays based on the same legend.[29]

Act I is set at the Japanese court in the early tenth century. Emperor Daigo, while touring the countryside, meets a destitute Princess Naohime, who is carrying her crying baby (whose father is Semimaru). The emperor takes pity and places them under his protection. In Scene ii, we meet Prince Semimaru's neglected and jealous wife. After sleeping once with Naohime a year earlier, Semimaru had fallen in love with her and consequently lost interest in his wife. He and Naohime secretly meet again at court, but are discovered by Hayahiro, the brother of Semimaru's wife. The wife becomes furious and sheds tears of blood. In the fantastic last scene, the wife visits a shrine of the god of jealousy. Here she meets another of Semimaru's lovers, Bashō; both vent curses at Naohime. The wife, excited with anger, turns into a vengeful serpent and disappears into the Uji River!

Act II is in the countryside where Semimaru and Naohime have fled, chased by the villainlike Hayahiro. Tadamitsu, a former court archer, saves the pair from Hayahiro's attack. Bashō, Tadamitsu's sister, comes seeking revenge against Semimaru, but is killed by Semimaru's servant Kiyotsura. The royal pair escape Hayahiro's attack again, but the villain kills Tadamitsu's elderly father.

In Act III, the now-blind Semimaru is sentenced by his father Emperor Daigo to banishment on Mt. Ausaka. In the second scene Semimaru journeys (*michiyuki*) to Ausaka where he is to be exiled. Naohime, who has fled to the mountains planning suicide, is restrained from killing herself by two mysterious temple boys, who take her to watch the suffering Semimaru but do not allow her to talk to him. In the final scene, Semimaru suffers alone and then experiences enlightenment, after which he and Naohime are reunited.

Act IV returns to the countryside where the persistent Hayahiro is in hot pursuit again, but he is driven away. Meanwhile, Semimaru and Naohime are hiding at the home of Hakuga, Semimaru's disciple in lute playing. Twice during this act, Naohime and Tadamitsu's old mother exchange places, which causes humorous incidents of mistaken identity. Hayahiro is once more chased away by Tadamitsu.

In Act V Tadamitsu, dressed as a priest while in pursuit of Hayahiro, the murderer of his father, recites a Buddhist sermon on the idea that enlightenment must be sought within one's own heart. He later finds and kills Hayahiro in an exciting battle. As scene ii unfolds Semimaru is staying at his sister's (Sakagami) mansion. Meanwhile, Naohime is pregnant and an exorcism ritual is to be held to calm Semimaru's dead wife's spirit—to allow for a safe birth and to restore Semimaru's sight. The last scene, the exorcism ritual performed by a Yamabushi priest, is a song. Finally, the spirit is pacified and becomes a Bodhisattva, an event which brings happiness to all.

If we look at the organization of acts and scenes (Chart 6), it becomes clear that, although the content is different from the other two plays analyzed, the musical and dramatic pattern is similar. In spatial terms, the play begins on a high plane in the emperor's presence and then moves to the Imperial Court. Act II shifts to the countryside where the action is performed by commoners. Act III, the scene of Semimaru's banishment, is spatially the farthest point away from the court, at Ausaka near a mountain pass. Act IV returns to the ordinary, middle world of the countryside after Semimaru's experience of enlightenment.

CHART 6
Outline of Five Acts of *Semimaru*

ACT I Setting: court, noble tone

i *jigoto*	ii *jigoto*	iii *jigoto*	iv *fushigoto*
Emperor's presence; meets Naohime	Emperor's court; Semimaru & wife	Semimaru and Naohime; discovery by wife	Hashihime Shrine; wife's jealousy and curse of death

ACT II Setting: countryside, active, violent

i *jigoto*	ii *jigoto*	iii *jigoto*
Village of Kohata; Tadamitsu saves Semimaru and Naohime from Hayahiro (villain)	Tadamitsu's home; Kiyotsura searches for Semimaru; Bashō comes home	Tadamitsu's home; Kiyotsura kills Bashō; Hayahiro attacks and kills Senju

ACT III Setting: deep in mountains, Semimaru's blindness and banishment

i *jigoto*	ii *fushigoto*	iii *mixture*	iv *jigoto*	v *fushigoto*
Emperor's court; blind Semimaru is banished to mountains	Semimaru's dark journey	Semimaru is abandoned	Naohime attempts suicide in despair; temple boys restrain her	Semimaru's suffering, alienation, and enlightenment

ACT IV Setting: countryside, active, comic

i *jigoto*	ii *jigoto*	iii *jigoto*	iv *jigoto*
Hayahiro in pursuit	Hakuga's cottage; Kiyotsura meets Naohime & both flee	Tadamitsu puts mother in cottage while fighting Hayahiro	Comic Hakuga & Semimaru think Naohime has become old woman; Tadamitsu chases Hayahiro

ACT V Setting: river crossing and court

i *mixture*	ii *jigoto*	iii *fushigoto*
Tadamitsu gives sermon on salvation then he kills father's enemy Hayahiro	Sakagami mansion; Sakagami calls for prayer to soothe Semimaru's wife	Uji River exorcism ritual; wife appears and becomes Kannon

The action of Act V returns to the higher aristocratic world, where the jealous spirit of Semimaru's wife is finally calmed through Buddhist rituals, allowing for a happy conclusion. Acts I, III, and V are related, and Acts II and IV are likewise complementary.

Chikamatsu, given the pattern of a five-act play outlined by his employers, gradually improves his skill in producing coherent narratives within the strictures of the conventions. Matisoff aptly remarks that the structure is "more tightly controlled than that of many of Chikamatsu's history plays," and suggests that it is closer to later *sewamono.*[30] In terms of time and space, *Semimaru* is short and narrow if it is compared to other history plays. Temporal and spatial unity, however, are less important than unity of theme. Love, jealousy, and poetry are the main themes, and the basic crisis occurs because of a love triangle—Semimaru, his wife, and his lover Naohime—a situation which later becomes central in many of Chikamatsu's contemporary-life plays. Semimaru's wife's curse, vented at the end of Act I as she dives into the Uji River at the darkest hour of the night and is transformed into an angry serpent, provokes the remaining action. Finally, during the Buddhist exorcism ritual in Act V, her angry spirit is calmed. These two scenes are songs. Both are musical flourishes and stand in contrast: the first is violent, downward, dark, and destructive; the second, calm, upward, bright, and creative. Although Matisoff contends that this final scene would not have been easily comprehensible or interesting for the audience,[31] there is evidence that it was a popular part of the play. Three of Gidayū's *danmonoshū, Tōryū jōruri kohyakuban* (One hundred favorites, 1703), *Jōruri kogiku maru* (Jōruri's small chrysanthemums, 1706), and *Ōmu ga soma* (Parrot Mountain, 1711) contain scenes from *Semimaru.* The first two have the Kibune Shrine scene and the final Buddhist exorcism ritual, while the last one contains these two and the third-act *michiyuki.* This indicates that in Chikamatsu's time there was interest in these three scenes. Although the audience might not have understood all the language in the Yamabushi ritual, they were probably fascinated by the mysterious ceremony and enjoyed the musical presentation of it.

Chikamatsu displays his developing skill as a poet and playwright by making the third act the central axis. *Semimaru* comes closer than the two previous works in meeting Gidayū's edict that this act contain the essence of the entire play. Semimaru, recently blinded from an eye disease, lives in darkness:

> The sun and moon in heaven have no light
> In black of night his lamp is dark.[32]

Semimaru is banished from the court and abandoned on Mt. Ausaka by an order from his "kind" father, Emperor Daigo, who hopes to help him in the next life by having Semimaru atone for his sins. He becomes a blind beggar (*yatsushi*) dressed in rags, his only belongings are but a lute and a bamboo staff. He has fallen from the high court to travel a lonely journey of suffering. For him, Mt. Ausaka, so far from civilization and the elegant life he knew, is a realm of alienation and hardship.

Act III is masterfully organized. According to the notation (*sanjū*), the act divides into five scenes. The Matisoff translation, however, has three scenes.[33] The first and fourth are *jigoto* dramatic scenes and the remaining three are *fushigoto* songs (iii is actually a mixture of styles). Another way to divide the act is into two parts: the first, *kuchi,* is the opening two scenes, and the second, *kiri,* is the last three scenes. Scene i is introductory to the *michiyuki* (ii), and there is an emotional climax in the *michiyuki.* The *michiyuki,* in turn, also sets the mood for the *kiri.* After the *michiyuki,* the tensions, that were pulled taut as Prince Semimaru made the journey into exile, relax. From Scene iii, two or more chanters take the various parts. After Semimaru is abandoned in Scene iii, the setting shifts abruptly to where Naohime is attempting to drown herself in a sacred stream at Ausaka. She is restrained by two temple boys, who then lead her to Semimaru but do not allow her to speak. She is forced to silently witness her lover's agonies.

It is difficult to imagine how this scene was staged then or might be staged today. The dramatic possibilities are fascinating. A different chanter represents each character, one for the blind, suffering prince, another for the mad lover near to, but unable to touch the man of her desire, and one for the two supernatural boys who direct the action. Scene v is the high point musically—or the low point spatially in that it takes place symbolically in hell—of the entire play. Let us look at the scene in detail.

Semimaru's lines begin a musical section leading to the climax of the play. The scene was sung by three chanters and contains various musical styles: Nō chanting and Heike singing, as well as others. The dramatic tension is maintained by the supernatural boys who keep Naohime from touching Semimaru. When Semimaru speaks he is alone:

He cannot see before his eyes
The foam of the rushing river
Nor understand the logic of this life,
So like that water foam.[34]

All of nature causes grief; it is the harsh agent of his suffering. Naohime too suffers:

Is my life a reflection in a mirror?
We meet and yet cannot speak.[35]

Semimaru refers again and again to his complete isolation from mankind. He plays his biwa and sings:

Wait, oh wait,
Wild geese of heaven
Bring me word
How is the autumn
In my home town?
Deep in the mountains
I live in loneliness
I have no friends
Beside my lute.[36]

The direction is further and further into alienation and misery. Rain falls harshly on him until even the temple boys take pity and extend a hat. He wonders why he is not getting wet, and then has a vision of his past glory at Court. The musical notation marks how the lines are delivered. The first half is sung to dance rhythms (to "walls hung with a rose of Sharon"),[37] and the second is in the Nō *kuse* style (to "inside my royal palanquin").[38] Since the *kuse* usually is found at the climax of a Nō play, the description of the vision is a musical flourish—a bright flash in a sea of darkness.

After this moment of bliss, Semimaru's mind returns to the

present and recalls Naohime. The memory of his former happiness, however, makes his present plight all the more bitter. His words of longing spur Naohime to try to call Semimaru, but she is restrained again and collapses. This sequence of Naohime's reactions is a primary unit, and the notation suggests that the moment of her collapse is a very high note (*ue*) of anguish, which is followed by the relaxing *fushi* cadence that ends the unit with the lines: "It is an impossible situation for an ordinary mortal."[39] This line is followed by a pause in the action, after which Semimaru suddenly casts aside his cherished lute and plectrum, and finally banishes Naohime from his mind. He abandons all desire. Into this context, Chikamatsu places Semimaru's famous poem on the Meeting Barrier, which captures his thoughts at that moment:

> This, now this,
> Where people come and people go
> Exchanging farewells
> For friends and strangers alike
> This is the Meeting Barrier.

Semimaru undergoes a pivotal change in attitude in which he accepts nature and man's position in it. Nature had seemed harsh and violent, now it is gentle:

> They part in the morning
> And meet in the evening
> At Mount Ausaka the traveler's journeys
> Are only the delusion of dreams
> If the rain will fall—let it fall!
> If the wind will blow—let it blow![40]

"Deep in the mountains is a fine place to live. Yes, now I am coming to understand the impermanence of life." After this moment of awakening, the tension is relaxed, and the temple boys allow the two lovers to be together.

Since this play was revived at least twice, the audience of Chikamatsu's time must have found the third act to be effective drama. If Semimaru's experience is presented well, the audience is able to participate in both the misery and the enlightened vision. Chikamatsu, like Zeami, organizes his drama to lead the audience to a new consciousness. The act is well structured musically with alternating movements that tighten and relax the tension and, imagistically, with the direction of the poetic journey moving at each stage further into darkness and solitude. The destructive movement downward into darkness, which began with Semimaru's wife's curse, has been halted. A bright, upward, and creative vision commences which will be fulfilled at the end. Chikamatsu, through the voices of the temple boys, expresses the significance of Semimaru's experience:

> The way to Buddhahood is not dark. The midnight of enlightenment leads to bright dawn. Though both your eyes are dark, the moon of your heart is bright. (First boy) I am Hitomaro. (Second boy) and I, Akahito. To teach the essence of poetry, our two spirits have manifested themselves. With us you will achieve deliverance to Buddhahood.[41]

However, the revengeful spirit of Semimaru's wife is about and will not be pacified until the exorcism ritual in Act V. This ritual is a symbolic journey from conception through gestation and on to birth (or rebirth), and the symbolism is the same as in the ascetic Yamabushi journey. The spirit eventually is freed from her passionate feelings of jealousy and reborn as a Bodhisattva, allowing for a happy conclusion.

THE BATTLES OF COXINGA

The structures of the above three plays, composed over a five-year span, are strikingly similar. Acts II and IV invariably stand apart in terms of atmosphere and social status of the characters. Each play contains a dialectic of opposing forces as well as a pattern of descending and ascending movements, revolving on suffering or death in the third or fourth acts. Chikamatsu molded his sources to fit the theatrical conventions. Although he continued to write for both Kaganojō and Gidayū, during the next decade (1690s) he wrote mostly for Kabuki. Not until after the success of his first *sewamono, Love Suicides at Sonezaki,* in 1703 did he return to fulltime writing for Jōruri, this time as the Takemoto-za staff playwright. The final history play we shall discuss was written after both Kaganojō and Gidayū had passed away.

Takemoto Gidayū's death in 1714 left the Takamoto-za without a master chanter as his successor. Though he had designated an heir—twenty-three-year-old Takemoto Masatayū (1691–1744), who later became Gidayū II and then Harima no shōjō, finally earning his right to be Gidayū's successor—in 1714, however, the Takemoto-za essentially had no star performer. Because of this, it was beginning to lose popularity to the rival Toyotake-za, which had the famous chanter Toyotake Wakatayū (1681–1764, later Echizen no shōjō), a former Gidayū disciple who had broken away to form his own troupe. In 1714 Wakatayū, at thirty-three, was just entering his prime. The choice of Masatayū as head chanter distressed some senior performers, who fled to the Toyotake-za, further eroding the Takemoto-za's drawing power. In this atmosphere of confusion, Chikamatsu was called upon to produce a play to restore the theater's fortunes. Chikamatsu was, at sixty-one, the senior member of the troupe. His *Kokusen'ya kassen* (The battles of Coxinga, 1715) fulfilled expectations better than anyone could have dreamed. It ran for seventeen months, a record unheard of at that time, and consequently became a classical model for later Jōruri playwrights.

With the challenge Chikamatsu faced, he might have attempted a strikingly new kind of Jōruri play. Although many exotic elements are in *The Battles of Coxinga,* the structure is founded almost exactly on the conventional outline established by Gidayū nearly thirty years earlier. Chikamatsu used this form to create a popular masterpiece of Japanese theater.

The scale of vision is far wider than most other Japanese works of literature or drama; the subject is the world, with the Chinese capital in the center at the highest point. However, although its scope is much grander than in the other three plays discussed, it does have fundamental similarities. Each act is distinct and self-contained. Act I in all cases is lofty in mood and filled with aristocrats of the ruling court. Acts II and IV contain the most action and movement, and all the plays end auspiciously. There is also the important balance between *jigoto* dramatic and *fushigoto* musical scenes. In *Coxinga,* the imaginative movement descends to the end of the third act, after which it ascends, ending at the original point of departure. Act III, as in all three previous plays, is solemn and contains suffering. We also see a dialectical structure of opposing forces. These are some obvious similarities. Though Chikamatsu stays within the bounds of these conventions, he is able to produce a more complicated and cohesive work.

During the thirty years between *The Soga Heir* and *The Battles of Coxinga* (1683 to 1715) Jōruri developed greatly. According to Yūda Yoshio, during this time, the practice of including comic interludes between acts was abandoned.[42] Mori Shū postulates that this must have occurred around 1706 because the five-act history play after this time doubled in size.[43] Plays were lengthened to fill the gaps, and therefore the number of chanters increased from two to four, to five, then six, or more, which meant that one act was no longer performed by one chanter; rather, several chanters performed each act. Therefore, Chikamatsu was encouraged or even forced to complicate the action and increase the number of characters.

The Battles of Coxinga is full of excitement. The formidable array of characters and actions is quite baffling at first reading. There are court intrigues, a battle with a tiger, ocean trips from Japan to China in magical boats, tragic suicides, enchanting Chinese princesses, Taoist immortals on mountain tops, and magic bridges that span mountain passes and then disintegrate in an instant under an enemy army's feet—just to mention a few elements of this full-day drama. Many readers must wonder how these elements could all fit together. Chikamatsu presents a wide variety of theatrical experiences, though few in an audience could appreciate all the complexity.[44] It is difficult to summarize succinctly a play that takes a full day to perform. The story takes place in mid-seventeenth-century China at the end of the Ming dynasty. Nanking, the capital, is the setting for the first and fifth acts.

In Act I, the evil minister Ri Tōten, through various machinations, betrays the emperor and sides with the Tartar generals to overthrow the Ming dynasty. Go Sankei is a loyal minister, whose wise counsel is ignored by the foolish emperor, and during the uprising he flees with the crown prince. The empire is thrust into chaos.

The setting of Act II is a quiet fishing village in southwest Japan where Ikkan, a loyal former Ming minister, is living after having been exiled many years earlier. He has a Japanese wife and a half-Chinese, half-Japanese son Watōnai, who is married to another Japanese woman, Komutsu. Princess Sendan miraculously arrives alone in a boat, which had floated from China after the insurrection. In a comic scene, Komutsu is at first jealous of this foreign beauty to whom Watōnai affectionately speaks in strange-sounding Chinese. After learning of the turmoil in China, Ikkan, Watōnai, and his mother set off by different routes to revive the Ming government. The final scene is Watōnai's battle in China with a giant tiger, which he conquers and makes an ally. His ferocious bravery convinces a Chinese army to join forces with him against the Tartars.

Act III takes place at Kanki's castle. He is the husband of Kinshōjo, the daughter of Ikkan and his former Chinese wife. Watōnai and Ikkan arrive and request that Kanki join forces with them to restore the Ming, but Kanki refuses because he does not want to appear weak by changing sides just to follow his wife's family. Caught in this dilemma, Kinshōjo kills herself to break the impasse, and Watōnai's mother follows her to death as well. After this tragedy Watōnai gains an ally, General Kanki, who promptly gives him the official title of "Coxinga."

Act IV opens with a *michiyuki* magical journey of Komutsu and Princess Sendan to China. Then the scene shifts to a mountain inhabited by Taoist immortals. From there, Go Sankei and the crown prince watch the action far below of Coxinga's battles against the Tartars. Go Sankei eventually joins Coxinga's army. Act V brings victory to Coxinga's offensive which had begun in Act IV. The Ming dynasty is restored (contrary to historical fact) and eternal peace proclaimed.

Let us first compare *Coxinga* with the precepts expounded by Gidayū. Gidayū puts the first act under the title, "Love." The opening scenes of *Coxinga* are filled with beautiful women, thousands of them, and the central crisis is sparked by the Tartar lord's desire for Empress Kasei. The notation for the "battle of flowers" second scene indicates that it is a *fushigoto* song. A pleasant appearance, created by the gorgeous ladies competing in a flower game, certainly fits Gidayū's prescription. The technique of opening on a high plane, shifting to a musical scene, and concluding with a vigorous and threatening scene is the pattern we have seen before. Gidayū says the final scene (*kiri*) of the first act must entice and excite the audience. In *Coxinga,* excitement abounds; the bright Ming empire falls into chaos as the northern Tartars overrun the capital.

Gidayū considered the opening lines of the play to be an important guide to the tone of the entire work. According to the notation, the prelude ends with the line, "In glorious Nanking where brilliance reigns."[45] These lines describe the palace as the "Gallery of Clouds," making it, imaginatively, the highest point in the world. Looking carefully, we notice that spring is passing as blossoms scatter, yet eternal spring is created by the thousands of colorfully attired women, whose flowered gowns and scented bodies evoke a lush atmosphere of paradise. In these few lines, Chikamatsu has also suggested that the palace is out of touch with the cycle of the seasons and out of contact with happenings in the real world. In short, time is out of order in the palace. The next unit reinforces these suggestions made in the opening lines: "The emperor delights in song, dancing, and revels, and within his lively palace he keeps three consorts, nine spouses of the second rank,"[46] and thousands of other women. The image is of one man, the emperor, surrounded by a horde of females, and in this world of pleasure, "in the dead of winter he is offered summer melons." Time and order are out of balance; the delicate façade of pleasant appearances totters precariously on the edge of chaos.

Chikamatsu is not only describing a corrupt, decadent court; it becomes clear that the natural forces of yin and yang are out of balance. The Chinese emperor is the Sun (and Son, to use a Shakespearean pun) of Heaven. The emperor as the sun figure is the yang element, yet he is surrounded by three thousand females, a situation not in balance even considering Chinese yin-yang theory in which the yang emperor is supposed to be surrounded by a substantial yin harem.

The organization is based on principles of dialectics and cyclical movement: one is the struggle of opposing forces (Tartar-Ming, feminine-masculine, and so forth), without which the Immortals on the mountain say there would be "no order in creation";[47] the second is the cyclical movement that makes a full circle, beginning and ending in the Nanking capital.

The other four acts, too, closely fit Gidayū's outline. The second act, Shura (battles), is comic and vigorously paced, with the battle of Watōnai (the future Coxinga) and the tiger as the focus. The third act is intensely tragic and contains the essence of the entire drama. The fourth has a colorful *michiyuki,* which leads the audience out of the dark pit of tragedy to the land of the Immortals on top of a magical mountain. The fifth act is auspicious and brings the play full circle, as Coxinga's forces march into Nanking, far up in the "Gallery of Clouds." The fact that Chikamatsu could follow so closely Gidayū's theoretical outline and produce his most famous play bears out the dramatic soundness of Takemoto Gidayū's formula.

In terms of spatial movement, the first act ends with a fall from the "Gallery of Clouds" into a lower world of chaos. In the second act, Watōnai travels with his mother from the "common" world of a Japanese fishing village across the sea to China downward "penetrating deeper into the forest,"[48] where Watōnai faces his first challenge, the Chinese tiger. In the third act, the setting is General Kanki's mighty castle, the Castle of Lions, described as a place of suffering, a hell of "absolute darkness" (*mumyō*, a Buddhist term for ignorance or delusion, literally meaning "no light," describes his castle); imaginatively, this is the darkest and deepest point away from the high and bright Nanking. The fourth act begins with a journey which leads the two women from the "common" Japanese village into the clouds over China, and the audience up from the "Castle of Absolute Darkness." The final act is an upward offensive leading back to the "Gallery of Clouds," the Nanking Palace, now an ordered world dominated by an army of males.

The timeless scene in the Land of the Immortals is a *fushigoto* song which provides a view of the entire play; again to borrow Gidayū's phrase, it is a still moment in a rushing stream. The *go* board, with its black and white stones, becomes an emblem both for the action of the play and for the raging battle below. The idea of a balance between the yin and yang forces, which the Immortals expound, suggests a more understandable meaning for the opening lines of the play. The yin (female) presence has overwhelmed the yang (male). The description of the palace projects it as a habitat of frivolity and desire, where the rational element (Go Sankei and Ikkan) has been banished. Throughout the action of the play, the victims of the fall into chaos are all females. By the time the action returns to Nanking, only one woman is left, Princess Sendan, who is more rational (yang) than her brother the frivolous emperor. The first victim is the Empress Kasei, next is the valiant wife of Go Sankei; and, finally, Kinshōjo and Watōnai's mother sacrifice themselves.

On the Mountain of the Immortals, when Go Sankei asks the old men what is the result of their contest of the *go* game (or symbolically of the opposing forces of yin and yang), they answer, "Does not the good or bad fortune of Mankind depend on the chance of the moment?"[49] This principle is a key to understanding the play, which begins in a completely yin (female) court and ends in a completely yang (male) dominated one. The movement is similar to the wheel of fortune in Western culture with its rising and falling action (although in this case there is a falling and rising pattern). The fortune of Watōnai's forces depends on whether General Kanki will become Watōnai's ally. Therefore, the crisis in his castle is the turning point. Until the end of Act III, the dark northern (yin) Tartars have dominated the southern (yang) Ming, but after the tragedy in the castle the pendulum of the "fortune of the moment" swings to Watōnai's favor. Time had been, until the tragedy in the castle, a devouring force (yin), but from the point when Watōnai is symbolically reborn through the sacrifice of others and receives his new name, Kokusen'ya (Coxinga) "Lord of the Imperial Surname," time becomes a positive force that brings victory and creates a new Nanking.

To understand the significance of the tragedy in the third act, the scene must be put in the context of the whole play. The play is ordered on a revolutionary principle, of which the ultimate goal is a balanced order represented by a harmonious court at Nanking. However, to recreate this order after a fall into chaos due to an imbalance of yin and yang forces, sacrifices are necessary. In the scheme of this play the female element (yin) is the victim. Seen in relation to the entire play, Kanki's initial refusal to join Watōnai—because his wife is a relative of Watōnai—makes sense, because, to return to order, all the yin elements must be subjected to the yang, which means society must become totally militaristic and devoid of "weak" human feelings (yin in relation to the rational element yang). This scene in Kanki's castle aptly illustrates Eliot's idea of "levels of significance." On the immediate level the two women, who voluntarily sacrifice themselves to save the nation, appear as paragons of loyalty; on a more subtle, abstract level, they represent imaginative forces within the yin-yang struggle for order.

The overall structure, however, is not more important than the action at any moment in performance. Each part of the play is carefully constructed as musical movements within the *jo-ha-kyū* pattern. Its primary units are passages that begin with *ji* or *ji iro* and end with *fushi* (*okuri,* or *sanjū*) (see glossaries). Act III, broken down according to the notation, contains twenty-five musical units (see Chart 7). The notation *sanjū* signals a change of scenery and is when chanters alternate in performance. Therefore, the act divides into two scenes: i (*kuchi*) with ten units, and ii (*kiri*) with fifteen. The divisions of each scene into five major movements have been made on the basis of content.

The act is intricately organized. The narrative technique is to speed up and slow down the tempo, to tighten and relax the tensions, and to alternate light comedy with serious intensity. Since Scene i is introductory to ii, we can expect that in i the intensities will build to a peak, but will primarily serve to lift the awareness and sensitivity of the audience in order to prepare them for the tragedy which follows. In Scene i movements A and B are introductory and set the background. Movement C is the emotional climax, and, within it, unit 6 is the point when the chanter displays his skill. This complicated unit can be further broken down into three musical movements: the first ends with the *suete* musical pattern, and the second concludes with a *fushi* cadence but is followed by a line of *kotoba,* which means that the unit continues. In Keene's translation the length of the unit is almost exactly two pages. In unit 6a Ikkan has his turn to earnestly plead that he is indeed Kinshōjo's long-lost father. In 6b the mood changes and the pace quickens as Kinshōjo asks for proof; the Chinese soldiers cry out as a chorus demanding proof of identity. The Chinese soldiers' voices are in song and the impatient voice of Watōnai is in *kotoba,* setting up the contrast of above and below.

Ikkan next explains in *kotoba* that he left a picture of himself with her. His voice changes to *ji iro haru* as he tells how old age has changed him. The next passage, beginning with Kinshōjo's belief that it is indeed her father, gradually develops musically into an emotional arialike song. First she uses a mirror to reflect Ikkan's face in order to compare it with the portrait, a histrionic technique, which in reality only makes the two objects more difficult to see. This passage is like a *fushigoto;* dramatic action ceases and lyrical song commands the stage. Through this rhetorical device, the audience suddenly is afforded a glimpse into the heart of Kinshōjo, a figure who has just appeared. After she compares the face and picture in the mirror, her voice continues entirely in *ji,* the most melodious delivery style. The pitch rises and falls with the fluctuations of emotion. Towards the end of the passage the high pitch *kin* is used three times to bring the scene to its pinnacle of intensity.

> The days and nights of twenty years have been hard to bear. How
> *kin*
> grateful I am that you have kindly stayed alive, and that I can behold
> *kin*
> my father! (Narrator) She weeps for joy, not caring who might hear.
> *u*
> Ikkan, choking with tears, clings to the gate tower. He looks up and
> *kin* *fushi* *haru*
> she looks down, their hearts too full for speech. Only their tears
>
> have no end.

The action is brought to a dramatic pose much like a Kabuki *mie,* in which movement stops and a tableau is created. This extremely long primary unit is followed by one of a mere three lines, which cushions the emotional flourish as it wanes, giving the audience a moment to relax before the action continues. The audience becomes emotionally involved with Kinshōjo and is prepared to see her as protagonist in the following scene.

Scene ii, beginning with a section of light comedy, clears the air in preparation for the intensities to follow; the audience is offered a respite in the midst of this dark, tragic act. The next significant action is Kanki's sudden attempt to kill his wife, a deed cursed by the mother as outrageous. The audience is shocked and made ready for Kinshōjo's suicide. The mother's pleas for her daughter's life in unit 17 comprise another emotional build-up, and the unit is musically complicated. Unit 23, where the dying Kinshōjo begs Kanki to join Watōnai, is the dramatic climax of this act and of the entire play. With her collapse in 23b, the play shifts direction as Kanki shows his

CHART 7

Musical Structure of *The Battles of Coxinga,* Act III[a]

i A Arrival at Kanki's castle

1. Father and son travel to Castle of Lions (ending with Keene, p. 229 "son-in-law")
2. a. High stone walls (*suete* p. 229 "in the sky")
   b. (*ji chū*) Fortress unmatched in Japan (p. 229 "in Japan")
3. Ikkan in dismay (p. 229 "murmurs unhappily")

B Decision to seek Kanki as ally

4. a. Watōnai's bravado vs. mother's caution (*fushi* p. 230 "enemy camp")
   b. (*ji iro*) Mother counsels to control emotions (*fushi* p. 230 "cautions him")
   c. (*kotoba*) Group asks for admittance (p. 231 "in tumult")

C Emotional meeting of father and daughter

5. Kinshōjo asks who it is (p. 231 "understandable")
6. a. Ikkan introduces himself as her father (*suete* p. 232 "persuasion")
   b. (*ji iro*) Kinshōjo asks for proof of parentage (*fushi* p. 232 "perilous")
   (b and c are emotional and musical flourishes)
   c. (*kotoba*) Reflection of portrait in mirror (p. 233 "no end") Peak of scene i.
7. Everyone: Watōnai and Chinese weep (p. 233 "overflowing tears")

D Mother is accepted if bound by ropes

8. Chinese stubbornly refuse entrance (p. 234 "despair")
9. a. Mother admitted if bound (p. 234 "bu on bu on")
   b. (*kotoba*) Mother calms Watōnai's anger (p. 235 "a Japanese")

E Water signal and entry into Gate of Absolute Darkness

10. White for yes and red for no: Gate of Absolute Darkness (*sanjū* p. 235)

ii A Comic scene with Chinese maidservants

11. Affection so strong between mother and Kinshōjo that no interpreter needed (p. 236 "or Japan")
12. Dutiful Kinshōjo (p. 236 "her solitude")
13. Maids' curiosity (p. 236 "and nod")
14. *Musubi* joke (p. 237 "her taste")

B Kanki's dilemma

15. a. Kanki's arrival (*suete* p. 237 "flesh and blood")
    b. (*kotoba*) Kanki wants to meet mother (*fushi* p. 238 "carried")
    c. (*chū*) Mother struggles out (p. 238 "with pity")
16. a. Mother transmits request of Watōnai (*suete* p. 239 "determination")
    b. (*ji haru*) Kanki attempts to kill Kinshōjo (p. 240 "choke with tears")

C Mother's defense of daughter
  17. a. Kanki explains action (p. 241 "around her")
    b. Mother's honor (*noru chū haru* p. 241 "different lights")
    c. (*ji haru*) Mercy of Japanese (p. 242 "Helpless tears")
  18. Kanki refuses Watōnai's request (p. 242 "her chamber")

D Kinshōjo's tragedy
  19. Mother is in despair (p. 242 "set flowing")
  20. Kinshōjo puts rouge in stream (p. 242 "yellow river")
  21. Watōnai angrily scales castle walls (p. 243 "Kanki's Castle")
  22. Kinshōjo's suicide is revealed (p. 243 "is stunned")
  23. a. Honorable and brave Kinshōjo (*fushi* p. 244 "wife's influence")
    b. (*kotoba*) Kinshōjo begs Kanki to join Watōnai (*noru* p. 244 "loses consciousness")
    c. (*ji chū*) Watōnai becomes Kokusen'ya (p. 244 "Kuai-chi mountain")

E Mother's sacrifice for the cause
  24. Mother and daughter die happily together (p. 245 "for this world")
  25. Coxinga is saving dragon

Note: [a]Two scenes: i and ii (division is based on *sanjū* notation). Capital letters mark movements within the scene (divisions based on content). Arabic numerals mark the primary units (*ji* to *fushi*), and small letters mark divisions within the units (based on the notation). Page references are to *Major Plays of Chikamatsu.*

resolve to join Watōnai in arms, and then ennobles him with a new name, Kokusen'ya (Coxinga). The play begins to move upward towards victory with Coxinga as the commanding general. Without the musical notation, it is difficult to perceive the rhetorical techniques the playwright and performers use to guide the audience to appreciate the intensities of the play.

Another approach is to analyze Chikamatsu's use of language in the various acts. As mentioned above, the opening of the play projects a vision of pleasure and frivolity, which is, however, only a delicate façade held up by a disintegrating order on the verge of collapse. Later in the act when Go Sankei rationally explains in terms of yin-yang theory, that Ri Tōten's trick of gouging out his left eye as a sign of apparent loyalty is, in reality, a signal for the Tartars to begin war against the Ming, the emperor replies, "Such logic could prove that snow was actually black!" Moments after the emperor's confused statement, the world literally falls into chaos as the Tartar invaders overthrow the capital. Keene notes that the quotations from classical

Figure 13. Ukiyoe Actor Print (Kunihisa, 1863) of The Battles of Coxinga *(Kokusen'ya kassen), Act III. Kinshōjo commits suicide, and her blood colors the stream red. At the right is Watōnai (Coxinga) (Kawarasaki Gonjūrō). Kinshōjo (Sawamura Tanosuke) and Kanki (Bandō Hikosaburō). Courtesy of Waseda Theater Museum.*

literature, which open the second and third acts, mean the opposite of the actual situation of Watōnai and Ikkan.[50] Hozumi Ikan, the Confucian scholar and friend of Chikamatsu, states in his *Naniwa miyage* (1738), that he could find no classical reference to the opening passage of Act III, and that it makes no sense.[51] These passages, seemingly out of place, suggest that the political chaos is also a state of linguistic confusion. Yet, Act III ends both with a sense of hope and a quotation, which for the first time fits the situation. Coxinga is the "dragon" who will restore order. To understand why this change occurs, we must look again at the action in Kanki's castle.

The main happenings in Act III are the suicides of Kinshōjo and Watōnai's mother. At the beginning Watōnai's forces approach Kinshōjo's husband, General Kanki, to ask him to join them against the Tartars. Since Kanki is initially absent, the castle guards allow only Watōnai's mother, bound with ropes, into the castle to meet Kinshōjo. The plan is that, if Kanki agrees to become Watōnai's ally, then white powder will be put into the stream flowing from the castle. If the answer is no, then rouge will be the signal. When Watōnai sees the red flow of water come out of the castle, he leaps over the wall to save his mother from the supposed enemy. Red means the opposite, however, because, since the crimson water is Kinshōjo's blood, Kanki may now become Watōnai's ally. Language until this point has been confused; here it means the opposite. Tragedy revolves on the paradox. After this incident Watōnai, formerly half-Japanese (*wa*) and half-Chinese (*tō*), becomes one as Kokusen'ya (Coxinga), the emblem of a nation.

Act IV shifts to the Mountain of the Immortals, who live "in perfect harmony" playing *go:*

> Their thoughts move freely, like a spider's thread in the air, and their bodies have become empty cocoons on a withered branch. This is the art of conversing by hand, far removed from the usages of the world.[52]

We have seen confused language in the world of chaos as well as the paradoxical language of opposites in the dark hell of delusion; now on this mountain the enlightened Immortals communicate without language. They converse by hand:

Figure 14. Scene from The Battles of Coxinga *(Kokusen'ya kassen), Act IV From an illustrated* Kokusen'ya kassen *(Battles of Coxinga). While Immortals play* go, *Go Sankei (holding crown prince) surveys the world below. Courtesy of Tokyo University Library.*

> If it looks like a *go* board to you, it is a *go* board, and for the eye that sees *go* stones, they are merely *go* stones. But there is a text which likens the world to a *go* board. For those who see with their minds, the center of the universe is here.[53]

On this mountain everyday language has been transcended.

Go Sankei mistakenly believes the vision is real and tries to touch the images before him, but the old men stay him, explaining in essence that all he has seen is simply a reflection "in the mirror of his loyal and faithful heart."[54] The men tell him:

> Both old men: Our home is the moon, where the leaves of the laurel blown over,
> Second: Appear to the eyes that see true the waxing
> First: And the waning, but
> Second: The ordinary man, confused of mind, takes it for a mere contest between *go* stones.
> First: The fish swimming in the water
> Second: Mistakes it for a fishhook.
> First: The bird soaring above the clouds
> Second: Is frightened, thinking it a bow.
> First: The moon does not descend.
> Second: Nor do the waters rise.[55]

The two men explain to Go Sankei that the moon of enlightenment will eventually shine brightly unhampered by any dark clouds, referring both to the future Ming victory and to enlightenment attainable by anyone. Go Sankei's interaction with the Immortals is a *fushigoto* scene. Therefore, it is a still, dreamlike moment when dramatic action ceases, and the audience is given a view from the *go* board of the revolving play, "a perspective that is and is not," in Shakespeare's phrase. In one sense, the two men are talking about the morally "right" forces of the loyal Ming adherents, able to "see" the magic bridge across the ravine, like the Israelites who could "see" the Red Sea open a path for them alone. But on another level of significance, the waxing and waning yin-yang elements are forces in a cosmological struggle that never ends (and will not end at the conclusion of the play), and in which the "good or bad fortune of mankind depends on the chance of the moment." The brilliant moonlight is what Eliot would call the meaning of the whole. Its significance emerges gradually after contemplation on the entire play. When the Ming forces are victorious at the end, we realize this has been achieved only after all the female (yin) elements have been sacrificed. The only two women who remain are Princess Sendan, who is more rational than her brother the former emperor, and Coxinga's wife, Komutsu, who upon hearing Watōnai had been given a new name as General Coxinga, "at once assumed male attire," and became another Ushiwaka, the child-name of the great General Yoshitsune. The ending, which returns to the Nanking Palace, is actually the exact opposite unbalanced situation of the beginning, because there is an army of men and one lone female (presumably a virgin), Princess Sendan, where there had originally been one emperor and a harem of thousands. Therefore, although Chikamatsu changed history to allow Coxinga a victory where he really met defeat, he has, nevertheless, by his ordering of the whole play made it clear that the fortune at the end is only a "chance of the moment," and that the light of the moon unfettered by dark clouds is there for the individual determined to see it.

The world of the history play is grand in scale and its range of action extends from the realistic level of human passion and pain to the heights of supernatural marvels. The drama is constructed to offer enough variety to keep an audience attentive for the expanse of an entire day. Due to the extremes in this variety, however, the plays, as dramatic literature, may seem at first reading to be complicated jumbles of incongruent materials. They are not. An awareness of the conventions of the tradition exposes the plays as well-ordered and purposeful dramas. Musical and dramatic orchestration of diverse acts and scenes are fundamental; the structure is founded on a principle of dialectics in continual motion. Opposing forces revolve in cyclical progression and carry the audience through the intensities of the drama, stimulating their imagination to perceive for a brief moment—in the midst of tragedy—insights into the human condition which ideally lead to the elusive goal of enlightenment. The fantastic circle of Jōruri drama forms a whole—a totality of theatrical experience exciting to behold.

CLEOPATRA: If it be love indeed, tell me how much.
ANTONY: There's beggery in the love that can be reckoned.
CLEOPATRA: I'll set a bourn how far to be beloved.
ANTONY: Then must thou needs find out new heaven, new earth.

Shakespeare, Antony and Cleopatra

. . . in Tragedy is embodied the eternal contradiction between man's weakness and his courage, his stupidity and his magnificence, his frailty and his strength.

F. L. Lucas, Tragedy

5 · *Descent to Paradise*

LOVE SUICIDES AT SONEZAKI

On the sixth day of the fourth month in 1703 the one-act play *Sonezaki shinjū* (Love suicides at Sonezaki) was performed at the Takemoto-za in Osaka. It was a tremendous success. The Takemoto-za, which had been in debt, suddenly became solvent and its star, Takemoto Gidayū, a widely popular entertainer. The characters were not historical personages, as in period plays, but folk from the streets of Osaka. Kabuki had long been portraying contemporary figures, and Chikamatsu himself had been writing Kabuki of this type for many years. But this was something dramatically new for the Jōruri stage. There are records that perhaps other *sewamono* (contemporary-life plays) were performed before *Sonezaki,* but their influence on the history of Jōruri is negligible. This work was epochal, because it ushered in a new age of popularity for Jōruri theater and introduced an exciting genre about contemporary life.

Love-suicide plays are usually based on real-life incidents of young couples who committed suicide. The basic theme is simple; the lovers, for various reasons, are unable to be together in this world as husband and wife and so choose to die together, believing explicitly that they will be reborn in paradise in perfect conjugal happiness. In *Sonezaki,* Tokubei, a young clerk in a soy-sauce shop, is in love with a courtesan, Ohatsu. He wishes to marry her but cannot afford to buy her contract. He foolishly loans his employer's money to a friend Kuheiji, but this "friend" betrays Tokubei and accuses him of stealing his seal and forging a promissory note. A desperate Tokubei comes to explain his plight to Ohatsu in the last half of the second scene. The pair decide to die together in honorable suicide, both to be together eternally and to show Osaka their pure and honest hearts. The final scene is the *michiyuki* journey to death.

Love Suicides at Sonezaki did not cover a whole day of performance. It was a *kiri,* the final piece in a day's program after a longer history play. Most later *sewamono* plays, which Chikamatsu or his rival Ki no Kaion wrote, were also performed as finales. They were composed as one-act (*dan*) dramas with three scenes (*maki*). In *Sonezaki,* this is obvious because there are no divisions in the published text. It is, however, divided into scenes by the musical notation *sanjū. Sewamono* after *Sonezaki* generally divide into three *maki;* in Chikamatsu's (and Ki no Kaion's) case the divisions are marked by *jo, chū, ge.* Nevertheless, each major scene is still denoted by the *sanjū* notation. The author's and performers' (as well as the audience's) conception of the play is that it is equivalent to one act of a five-act history play. The early book on Jōruri, *Imamukashi ayatsuri nendaiki* (A chronology of the puppet theater, 1727), makes this clear: "An incident of two lovers committing suicide at the Tenjin Shrine in Sonezaki . . . was máde into a one-*dan* Jōruri and titled *Sonezaki shinjū.*"[1] In modern times, Kuroki Kanzō and others have also considered the *sewamono* equivalent to the third act of a five-act history play.[2] If we accept this notion, then the entire *sewamono* lies within the dark lower half of the full-day, history-play cycle. This one-*dan*:five-*dan* contrast is a fundamental distinction, because the conventions and consequent expectations of the performers and audience are different for each type of play. The *sewamono* is a single, complete unit, which maintains a uniform tone throughout and is, therefore, closer to a single Nō play. *Jidaimono* and *sewamono* plays must be compared, criticized, and appreciated with this distinction in mind.

Characters in most love-suicide plays are the lowest members of society; many male protagonists are orphans, and the females, who are low-level courtesans, come from desperately poor families. Often the girl has only a destitute mother, who ekes out a meager existence by some menial work such as mending clothes in a back alley. Furthermore, courtesans were officially treated as "outcasts." The characters have no court or high position to fall from, nor return to, and the grand world of public power or government is far removed. *Love Suicides at Sonezaki,* like most love-suicide plays, begins with the main characters down on their luck.

In a history play the setting is always noble and usually in a court at the beginning and end. Also, the characters who dominate the action at the beginning and end are usually aristocrats. In the Kabuki *yatsushi* play, as well, the main characters are of the upper classes. They may suffer agonies in the course of the drama but in the end will inevitably return to their former positions above the dirt and pain. Often in the third act of a history play the characters who sacrifice their lives are not the figures of nobility who began the play. These protagonists are often commoners. Two famous post-Chikamatsu examples, *Yoshitsune sembon zakura* (Yoshitsune and the thousand cherry trees, 1747) and *Kanadehon chūshingura* (The treasury of loyal retainers, 1748), in the third acts, both contain tragedies of characters not directly involved in the noble happenings which began the plays. Gonta and his family in *The Thousand Cherry Trees* and Okaru and Kampei in *The Treasury* are tragic figures of low social standing. In contrast, the love-suicide play begins and ends in the lower world of pain and suffering, from which the only escape is the dream vision of being reborn together on a lotus calyx in the next life. In general, characters who suffer tragedy in Jōruri are not lords and princes, as in Western drama; rather they are frequently common folk. It appears to have been a convention, both of life and theater, that the upper classes of men or the top ranks of courtesans never committed love suicide. Love suicide is only for the outcasts of society with no place of refuge.

Love Suicides at Sonezaki is compact in terms of temporal movement. It begins in daylight, continues into evening, and ends at the breaking of dawn. Though later love-suicide plays have more complicated plots, movement is always downward into the dark night of suffering from which the only release is death at dawn. The *michiyuki* journey to death is the crucial part of the poetic movement of the play, although, in terms of dramatic progression, it is hardly important at all. Rather, the *michiyuki* travel song is the quiet pool in the stream through which, as Keene has suggested, the characters grow to heroic stature.[3] During the journey the characters painfully recall loved ones whom they are leaving behind and are forced to reflect on the problems which brought them to this impasse. The *michiyuki* is a compressed vision of crucial episodes in their lives and is therefore a summation of the whole play. There are no new dramatic developments, but the audience watches the characters mature spiritually in the face of imminent death.

Yūda, in an article published in 1964, analyzed the *michiyuki* of *Sonezaki* as the five-movement pattern of *jo-ha-kyū*.[4] His posthumously published (1977) collection of seven Chikamatsu plays also contains a breakdown of this *michiyuki* into five movements, but he makes some divisions at different points.[5] As he examined other *michiyuki* passages and discovered similar patterns of literary and musical conventions in later *sewamono* plays, he seems to have altered his theories. He analyzed the basic pattern of love-suicide *michiyuki* into a five-part sequence:

A. *Makura* (introduction)
B. *Kudoki* (regrets) or departure
C. Song or *michiyuki*
D. Song or dance; or Buddhist chant
E. *Kiri* (finale)

Order of middle three varies in each play

Scholars still debate where the divisions occur in the *michiyuki*, because there are not always clear signals as there are for primary units. Sometimes the musical notation clearly marks a division, but often one must rely on content as the determining factor. After examining *michiyuki* that Yūda did not analyze, I am convinced that there is a pattern of five movements. Usually the central part is the *kudoki*, where the pair take turns grieving over their fate, which is both to suffer and to cause suffering. The *kiri*, however, is usually an emotional uplift as the pair entrust their souls to Buddha and experience a vision of happiness in the afterlife.

Yūda's division of the *michiyuki* of *Sonezaki* into five movements is as follows. (Line references are to Keene, *Major Plays of Chikamatsu*.)

A. *Makura*	to "Dream of a dream" (p. 51)
B. Umeda (Magpie) Bridge	to "tears they shed" (p. 52)
C. Jōruri song (*kudoki*)	to "leads her by the hand" (p. 53)
D. Umeda Embankment	to "tears increase the sum" (p. 54)
E. *Kiri,* Arrival	to "thick wood of Sonezaki" (p. 54)

The *makura* is a poetic invocation of the lovers' sad plight. In movement B, they vow to be husband and wife for eternity. The metaphor of the Umeda Bridge symbolically becomes the famous magpie bridge over the Milky Way, which joins for one day a year the stars of the Weaver girl and the Oxherd boy in the Chinese legend. In movement C, Chikamatsu uses the partying folk, who drunkenly sing of love suicides, as a technique of expressing feelings through an outside voice. In this song, the pair symbolically die. Movement D begins with a reference to their dead bodies being ravens' food, but then it quickly rises in a flourish of hope as the lovers imagine rebirth on the same lotus. The final movement is calm after the frenzy of the middle three sections in which they imagined in quick succession their deaths, their decaying bodies, and their rebirth into a happy world. The lovers arrive at the chosen site and never once look back at the world they leave behind. Tokubei and Ohatsu, of all Chikamatsu's *sewamono* heroes and heroines, are the most naive, pure, and guiltless. For them, death is a beautiful release from troubled lives. During the journey their minds are fixed upon a vision of a future union in bliss atop the lotus flower. This simple, innocent joy in death is not found in its fullness again in Chikamatsu's love-suicide plays. The road to paradise becomes increasingly more rocky and difficult.

In *Love Suicides at Sonezaki,* as the lovers set out on their journey to death, they are full of joyful expectation. "They exchange glances and cry out for joy, happy that they are to die." The characters in the later *Shinjū nimai ezōshi* (Love suicides and double-folded picture books, 1706), however, are never allowed to be together throughout the play, and, since they are unable to die together, they kill themselves at separate locations, after counting 10,000 beads on rosaries, in order to die at the same moment. In Chikamatsu's next play written the same year, *Hijirimen uzuki no momiji* (Scarlet crepe and April autumn color, 1706),[6] Yohei fails to die after killing his wife Okame; he is unwillingly saved and forced to live on to pray for his wife's soul. Death is not easy nor joyful as for Ohatsu and Tokubei, who in contrast seem to have rashly chosen suicide without considering the consequences. This difference is reflected in the *michiyuki* of these two later plays.

In *Double-folded Picture Books,* just before the *michiyuki* begins, the final words of the previous scene express the idea that the spirits of the lovers are leaving their bodies and will be joined after death. Although this play contains a journey to death, it is a spiritual rather than physical journey. The title of the *michiyuki* is "Chishigo no michiyuki," a journey at the moment of death. Movement A refers to the rosary beads they will count as they prepare for death. Then follows mention of hell and the 108 sufferings they will encounter.[7] In movement B, the dominant image is a falling star, a symbol for their spirits traveling to join each other. The lovers hug each other, but in the next instant the action shifts to a graveyard, and the movement ends with rainfall in the dark night.[8] In movement C, Ichiroemon recalls *Love Suicides at Sonezaki,* and then Oshima far away does as well. Oshima considers herself beastly to cause Ichiroemon's death. The imagery projects a downward movement ("bury their name in mud" and "fall down into the River of Hell"). The reference to the River of Hell divides the section into two parts. The imagery, which had been attributed to Oshima, switches to depict Ichiroemon's actions as he wanders lost in the darkness of delusion. Then both realize that their shadows, his in the moonlight and hers in candlelight, have disappeared; the pair are already dead.[9] Finally in the *kiri,* the voices become joyous with the realization they will die together, "separate but as reflections in a mirror," and the first glimmer of dawn breaks through the darkness. During their journey the lovers first descend into the miseries of hell, where they regret their actions and then imagine themselves already dead. After the moment of darkest gloom, however, love lifts them upward, bringing the journey to the joy of a new dawn.

The *michiyuki* in *Scarlet Crepe and April Autumn Color* can also be divided into five movements, which are outlined below, in this case using the analysis of Yūda.[10]

Figure 15. Scenes from Love Suicides and Double-folded Picture Books *(Shinjū nimai ezōshi). From* Eiri-jōruri *picture book. The scene at the top shows Ichiroemon's frustration when he fails to see Oshima who has gone on a boating excursion with a guest. The lower scene depicts the pair dying separately, but their hearts traveling to be united. Courtesy of Shōgakkan.*

A. Fear of a wild dog; messenger from hell waits for them.
B. Reference to River of Death, drum strikes fear of Shura realm.
C. Okame's *kudoki;* she expresses regrets about leaving the world; reference to Teika poem about burning body and salt fires.
D. Reference to *Yaoya Oshichi,* the story of a young girl who is burned at the stake for her pursuit of love; the lovers imagine how their death will be seed for gossip and popular picture books. They weep at the thought of their own cremation. Grasses without roots become grasses burned in the fires.
E. Dawn begins to break; they must hide from early-rising farmers and fishermen.

The pattern is similar to *Double-folded Picture Books* in that the second, third, and fourth movements constitute a symbolic journey through death and the tortures of hell. In all three *michiyuki* examined, the pair has, through poetry and song, symbolically experienced death and journeyed through the lower realms of the Buddhist Six Paths. In *Double-folded Picture Books,* Oshima considers herself a beast, and in *Scarlet Crepe and April Autumn Color* the couple experience the terror of the Shura realm. Each time they must witness the physical death and decay of their bodies.

The plots of these two plays are similar to *Love Suicides at Sonezaki,* because the basic conflicts depend on the machinations of a villain. Each pair of lovers, consequently, appear to be innocent victims of circumstances beyond their control. However, *Shinjū kasane isutzu* (Love suicides at the sunken well, 1707) is different and heralds an important development in the love-suicide genre. There is no villain, and the situation is antecedent to the later masterpiece, *Shinjū ten no Amijima* (Love suicides at Amijima, 1721), because in the *Sunken Well* the male lead Tokubei has a wife and two children whom he abandons to commit suicide with a courtesan, Ofusa. In the *Sunken Well,* Chikamatsu sets up an opposition between *giri* (moral responsibility, reason) and *ninjō* (human love or desire). One manifestation of this opposition is the different settings in each of the scenes (*maki*). Scene i takes place in the dyer's shop, which the wife has inherited; here Tokubei lives with his family. The shop-home is the sphere of *giri,* where Tokubei has family and business responsibilities. In contrast, the setting of Scene ii is the pleasure quarter, where natural or animalistic desires are unhampered by the strict societal codes. Tokubei is the first of Chikamatsu's characters to stumble confused between these two worlds and encounter no villains except the one within himself.

Chikamatsu has Tokubei face the weight of moral responsibility in each scene. In the first, the father-in-law's harsh words of reproof and his wife Otatsu's pleas shame Tokubei to the point where he admits his guilt and apologizes for ruining the family. The scene ends with confusion between duty and desire. "His heart is split in two"; he finally chooses desire and flees to the pleasure quarter. In Scene ii, Tokubei's brother and sister-in-law, who are proprietors of a teahouse, counsel Tokubei and Ofusa to dissuade them from suicide. Tokubei is again shamed, and the climax is the torturing of Ofusa, who is hiding beneath the *kotatsu* heater, by the brother who pours hot coals into it, knowing she is hiding there. Finally, in the last scene, just before the lovers are to commit suicide, Otatsu comes running with the children after Tokubei, begging the pair not to die and leave her and the children alone. To the bitter end, Tokubei is forced again and again to reject his wife and family for the sake of passion.

Yokoyama Tadashi has proposed a method of interpreting Chikamatsu's love-suicide *michiyuki.*[11] He found in Chikamatsu's later love-suicide plays, from the *Sunken Well* (1707) onward, that *michiyuki* scenes are "compact retrospective summations" of the central problems of the entire play. This is an interesting view of the function of *michiyuki* in love-suicide plays. Although the tendency to become more of a retrospective summation is evident in plays after the *Sunken Well,* the pattern is basically the same as in the three plays discussed above. In the first movement, the pair travel from "this shore" (this world) to the "other shore" (Buddhist paradise) on a fleeting bridge of reality ("shibashi kono kishi kano kishi no / kari no utsutsu no karibashi ya").[12] This suggests the journey is from this life to the next, and therefore that the spirits are no longer of this world. In the second movement, the pair travel through the

Dōtombori theater district and imagine their love suicide being enacted on the stage. The dominant image is the "darkness of love," and the pair see themselves as following in the tradition of *Love Suicides at Sonezaki.* In the fourth movement, voices call out for them, and they strive to flee the "shark jaws" of this world. The Buddhist assertion that sin and suffering can lead to salvation is their guiding phrase, and they begin chanting to Buddha. In the final *kiri,* there are the words, "morning was this world; night is the next; we are in between. Tonight we will become one." Movement is first downward into pain and darkness, but changes upward towards hope in the fourth and fifth movements, when the lovers call out to Buddha.

This *michiyuki* states explicitly that the lovers are traveling from this life to the next, and we have seen how in every example examined the pair symbolically die. Chikamatsu makes the love-suicide *michiyuki* a spiritual journey through the tortures of the Buddhist lower realms. Are the lovers already dead when they set out on their journey? Gunji Masakatsu has suggested such an interpretation.[13] He points to the folk tradition of pilgrimage journeys to support his idea that during the *michiyuki* the characters die, travel through the nether world, and then are finally prepared to die physically and obtain salvation. He cites the popular pilgrimage to the Kumano Shrines as an example of a journey to the other world of death followed by a return to this life. The Kumano Shrines are among the most important mountain shrine-temples of the Yamabushi tradition, and we have seen in chapter 4 that the Yamabushi pilgrimage to such sacred mountains is considered a symbolic journey through the Six Realms and rebirth from the womb. The journey frees one from the never ending cycle of death and rebirth, allowing one to achieve Buddhahood.

Chikamatsu invariably suggests that the lovers travel through the sufferings of the nether world. In *Shinjū mannensō* (Love suicides at the women's temple, 1710), the lovers set out along "paths (which) are many at the crossings of the Six Ways,"[14] and the last line of the opening of the *michiyuki* is "the tears they shed flow into each other's mouths, like water to moisten the lips of the dying." Just a few moments after this *michiyuki* Kumenosuke's sister soaks the stonecrop plant which withers, a sign that Kumenosuke is already dead. In *Tamba Yosaku matsuyo no Komuro-bushi* (Yosaku from Tamba, 1708), which is not a love-suicide play but which has a love-suicide *michiyuki,* the pair travel explicitly along the Six Paths. In *Imamiya shinjū* (Love suicides at Imamiya, 1711) and *Ikudama shinjū* (Love suicides at Ikudama, 1715), there are similar references to traveling across the River of Death. In *Love Suicides at Amijima* (1721), the couple cross six bridges on their journey from this world to the Western Paradise and suffer torments of hell along the way. If Yokoyama's interpretation of the *michiyuki* as being a microcosm of the whole play is accurate, then the entire play can be viewed as a journey through the agonies of the lower samsaric realms. Examination of additional plays reveals this to be a meaningful interpretation.

Of Chikamatsu's twenty-four *sewamono,* eleven are love-suicide plays. The form is quite simple; there are two general patterns. *Love Suicides at Sonezaki* is representative of the first, and *Love Suicides at Amijima* of the second. The structure is based on musical patterns similar to those in history plays. Yūda has determined that each *maki* (scene) contains, as a rule, five movements divided by content. A 5-5-3 pattern is found in most later love-suicide plays as well: *Scarlet Crepe and April Autumn Color* (1706), *Love Suicides at the Sunken Well* (1707), *Love Suicides in Midsummer with an Icy Blade* (1709), *Love Suicides at the Women's Temple* (1710), *Love Suicides at Imamiya* (1711), and *Love Suicides at Ikudama* (1715). The rest have a 5-5-6 pattern. Ki no Kaion's various love-suicide plays utilize both of these patterns. These movements are organized like history plays on the *jo-ha-kyū* musical principle and consist of a number of primary units. An understanding of the intricacies of the narrative structure elucidates Chikamatsu's methods of leading his audience and readers into the intensities of tragedy. To explain in more concrete terms how the *maki* are organized, let us examine the second one in *Love Suicides at Sonezaki.*

Both movements and primary units entice the audience further into the dramatic action by bringing the tempo and tension to a peak, and then relaxing these before continuing. Movement A, beginning with a lively song, creates the atmosphere of the pleasure quarter. Next we meet a weeping Ohatsu. The juxtapositions accentuate the isolation of Ohatsu from the

CHART 8
Outline of Three *Maki* of *Love Suicides at Sonezaki*[a]

Maki i: *jo*
- A. Pilgrimage to 33 Kannon temples (*fushigoto,* not translated in Keene)
- B. Tokubei's explanation of his plight (to "Don't worry. Oh—look there, Ohatsu!" p. 43)
- C. Kuheiji's trickery (to "extraordinary display of feigned innocence." p. 45)
- D. Fight (to "Who beats him? There is no way to tell." p. 46)
- E. Decision by Tokubei to die to prove his honor

Maki ii: *ha*
- A. Ohatsu's troubled heart (to "I wish I were dead already. She can only weep." p. 48)
- B. Tokubei explains his woes; Tokubei moves under veranda (to "She assumes an air of unconcern." p. 49)
- C. Decision of lovers to die (to "they noisily depart." p. 50)
- D. Farewell to all (to "two o'clock in the morning has come." pp. 50–51)
- E. Escape from the house

Maki iii: *kyū*
- A. *Michiyuki* (to "They have reached the thick wood of Sonezaki." p. 54)
- B. Recollection, regrets, and farewell (to "shed incessant tears in all too understandable emotion." p. 56)
- C. Final suicide

Note: [a]The page references are to *Major Plays of Chikamatsu.*

bright world of fun and pleasure. The other courtesan's words in the following unit cut deeply into her heart, and in unit 4 she wishes for death. The movement has contrasts and parallels. The first and third units are lively with several characters singing and talking, and contrast effectively with the quiet second and fourth units, which focus on a single character—Ohatsu. We must quickly be coaxed into sympathy for our lonely heroine, and Chikamatsu gives the chanter subtle means to effect that end.

Tokubei's arrival, in movement B, gives an immediate uplift to Ohatsu, even though she is trapped and cannot at first go to him. In movement A, the spotlight was on Ohatsu; in movement B, it is on Tokubei. He explains his predicament and resolve to die; each alone makes the fatal decision. In movement C, their resolution to die separately becomes a resolution to die together. The dramatic build-up to this decision leads to the climax of

CHART 9
Musical structure of *Love Suicides at Sonezaki, Maki* ii[a]

A. Ohatsu's Troubled Heart (*Jo*)
  1. Lively song description of pleasure quarter (to "What a lively place this New Quarter is!" p. 47)
  2. Ohatsu's anxious heart; she weeps ("other friends come for a little chat." p. 47)
  3. Courtesans discuss rumors of Tokubei's shame ("Sympathy makes their visit the more painful." p. 48)
  4. Ohatsu tells them to stop talking; she wishes to be dead ("I wish I were dead already." p. 48)

B. Tokubei explains his woes; Tokubei moves under veranda (*Ha*)
  5. a. Ohatsu notices Tokubei, but cannot go to him ("she cannot do as she pleases." p. 48)
     b. Ohatsu goes out to Tokubei ("weeps in secret, soundless, painful tears." p. 48)
  6. Tokubei explains the trickery of Kuheiji, his shame, and resolve to die ("he slips beneath the porch at the step." pp. 48–49)
  7. He crawls beneath veranda ("She assumes an air of unconcern." p. 49)

C. Decision of lovers to die (*Ha*$_2$)
  8. Kuheiji tells lies about Tokubei ("He pours out his words convincingly." p. 49)
  9. Tokubei trembles with rage ("he leaves the room," p. 49)
  10. a. Ohatsu asks Tokubei if he wishes to die; he passes her foot across his neck ("to let her know that he is bent on suicide." p. 49)
      b. Bittersweet joy in decision to die ("they silently weep," p. 50)
  11. Kuheiji leaves ("they noisily depart," p. 50)

D. Farewell to all (*Ha*$_3$)
  12. Ohatsu quietly bids farewell to all ("men who do not realize the truth in time!" p. 50)
  13. Shop is closed for the night ("two o'clock in the morning has come." p. 50)

E. Escape (*Kyū*)
  14. Preparation to leave; extinguish light ("the lovers grope for each other—a fearful moment." p. 51)
  15. They open the door to rhythm of maid striking flint

Note: [a]Capital letters mark five movements; arabic numerals designate primary units; and small letters, subdivisions within primary units. Page references are to *Major Plays of Chikamatsu*.

movement C, of the second scene, and of the entire play. The choice is bittersweet. Yet, at this dramatic moment, serious drama could turn to farce if the tension is not kept taut, because the action of Tokubei—pulling the puppet's foot across his neck as a sign that he will join her in death—can destroy the delicate line between illusion and reality upon which the puppet theater depends.

Let us look more closely at this passage. The following is based on Keene's translation, and is unit 10 in the above outline. Ohatsu is speaking both to the villain Kuheiji and to Tokubei, who is hiding under the veranda. (See glossaries.)

ji iro haru u iro
a. Ohatsu weeping bitterly, exclaims, "You needn't try clever words on
kotoba
me. Tokubei and I have been intimate for years. We've told each
other our inmost secrets. He hasn't a particle of deceit in him, the
ji iro u u
poor boy. His generosity has been his undoing. He's been tricked,
haru
but he hasn't the evidence to prove it. After what has happened
u
Tokubei has no choice but to kill himself. I wish I knew whether or
u
not he was resolved to die." She pretends to be talking to herself,
u
but with her foot she questions him. He nods, and taking her ankle,
u fushi
passes it across his throat, to let her know that he is bent on suicide.

chū haru iro kotoba
b. "I knew it. I knew it. No matter how long one lives, it comes to the
ji haru iro
same thing. Only death can wipe out the disgrace." Kuheiji is startled
by her words. "What is Ohatsu talking about? Why should Tokubei
kill himself? Well, if he kills himself, I'll take good care of you after
he's gone! I think you've fallen for me too!"

"That's most generous of you, I'm sure. But would you object if, by
ji haru u
way of thanks for your kindness, I killed you? Could I go on living
u u iro kotoba
even for a moment if separated from Toku? Kuheiji, you dirty thief!
ji haru
Anyone hearing your silly lies can only suspect you. I'm sure that

u *ue*
Toku intends to die with me, as I with him." She taps her foot, and

Tokubei, weeping, takes it in his hands and reverently touches it to
u *u* *u*
his forehead. He embraces her knees and sheds tears of love. She too
u *u*
can hardly conceal her emotions. Though no word is spoken, answer-
kin *fushi chū* *haru*
ing each other heart to heart, they silently weep.

The unit divides into two halves, according to the notation (*fushi* followed by *chū*). In the first part Ohatsu adroitly asks Tokubei if he will die with her. Tokubei's reply ends this first part. Because of the musical arrangement, the first section is necessarily an introduction to the second half, and, although the first half is a crucial point in the drama, the second part is the musical climax. After Tokubei's affirmation of love, tension rises and the tempo quickens as Ohatsu's voice displays her growing confidence; she becomes openly defiant toward Kuheiji. She is excited, and her words, about life coming to the same thing in the end, are rapid lines of *kotoba,* which contrast with the preceding slower-paced song passage. The line, "only death can wipe out the disgrace," is delivered in the adorned *ji* style with a high pitch. This line stands alone with added significance, because the styles preceding and following are in cool *kotoba.* Passages spoken by and to Kuheiji are in *iro,* a lively dramatic delivery style in between speech and song, in this case the voice of a crafty villain. Ohatsu answers him in this same style with defiance and sarcasm but, when thinking of living without Tokubei, her voice alters to a high-pitched song style to heighten the effect. When speaking to Kuheiji, however, her voice again changes to cool speech, as she upbraids him for his dirty tricks. Then, when directed towards Tokubei again, her voice returns to the "aria" style, and her monologue ends with a musical flourish. After this, the spotlight narrows on the two lovers alone as they silently weep. The final part has an effect like an aria in drawing the audience into sympathy for the lovers. The stage arrangement encourages this as well, because we are party to the secret communication of the couple.

The atmosphere suddenly alters after the decision is made.

Once death is chosen, ironically, the tension dissipates, and movements D and E are almost joyful at the prospect of death. The lovers' escape has its moments of suspense but comedy reigns, as we watch the naked maid blindly scurrying about trying

Figure 16. Scene from Love Suicides at Sonezaki.
From cover of Sonezaki moyō *(Patterns of Sonezaki, 1761),*
a jōruri ezukushi *picture book for a performance in the Toyotake theater.*
Courtesy of Waseda Theater Museum.

to light the lamp Ohatsu has extinguished. However, there is no doubt they will die. Their exit is the *kyū*, quick-paced conclusion, and a *danmari* scene of stylized dance movements in the dark. Looking again at the entire play, we see that the *jo-ha-kyū* pattern accurately describes its structure. In *maki* i, the problem evolves; in ii, the crisis and consequent decision to die occur; and in iii, the decision of *maki* ii is completed during the final dance of death. The form is simple and the central characters ingenuous; qualities that have perhaps fostered the play's enduring success.

The novelty and subsequent popularity of this play made it the model for later variations on the love-suicide theme. In *Sonezaki,* the couple is a young clerk and a young courtesan, but, in later plays, there are some differences in character types. In *Scarlet Crepe and April Autumn Color,* the pair is a young married couple driven out of their home by a villainous stepmother and her son. Chikamatsu's last *sewamono, Shinjū yoigōshin* (Love suicides on the eve of the Kōshin Festival, 1722), also has a married couple as the main characters. *Love Suicides at the Women's Temple* depicts the somewhat unusual situation of a young monk and a merchant's daughter committing a love suicide. Chikamatsu's *Love Suicides at Imamiya* and Kaion's *Imamiya shinjū marukoshi renrimatsu* (Love suicides at Imamiya near a double pine tree, 1711) and *Tamoto no shirashibori* (Sleeves wet with tears, 1710; Osome and Hisamatsu story), all have male shop clerks who fall in love with other workers, or, as in *Sleeves Wet with Tears,* with Osome the owner's only daughter. In all these cases, however, the plot is generally only a variation on the *Sonezaki* pattern.

LOVE SUICIDES AT THE WOMEN'S TEMPLE

Love Suicides at the Women's Temple (1710) is unusual because the lovers are not social outcasts but are cast out by their families; Kumenosuke, a young Shingon Buddhist monk, lives at a monastery on Mt. Kōya; Oume, the only daughter of a relatively prosperous merchant family, lives nearby. *Maki* i, set on Mt. Kōya, introduces the monks' world where pederasty is the way of life. Oume has sent two letters—one to the high priest and the other to Kumenosuke. The first letter is a forgery, stating that Kumenosuke's father wants him to return home. The second is a love letter to Kumenosuke, explaining her plan. However, the letters mistakenly reach the wrong people. Meanwhile, the elder brother of a boy that Kumenosuke accidently killed many years earlier arrives at the temple. He is angry that Kumenosuke wishes to leave the priesthood. During their conversation, the high priest enters and denounces Kumenosuke for polluting the mountain sanctuary (only relations between men are allowed) with his love for a young girl. After being abused verbally and physically, Kumenosuke is literally thrown off the mountain, and even the heavens are angered and spit down wind, rain, and hail in a furious storm—all for the love of a young woman.

Maki ii takes place in Oume's home, where the parents are planning to have their daughter marry a rich Kyoto merchant—against her wishes. Kumenosuke, in tatters, arrives; Oume shelters him in the house much to her parents' chagrin. At that moment, the groom enters furious after hearing rumors of Oume's love affair, and demands return of the wedding money. Oume's father and the groom actually come to blows before the mother steps in to settle things. She pleads for calm, speaking both to the men downstairs and to the lovers upstairs. Finally Oume emerges—completely unrepentent—but the groom is nevertheless charmed and forgives her. He immediately follows her upstairs into the room where Kumenosuke is still hiding. The bedroom scene is suspenseful but borders on the comic, as the groom is made the fool. Finally, in darkness, Oume's mother leads Kumenosuke out of the house, unaware that Oume is following.

The final *maki* is the journey to death on Mt. Kōya. The pair encounter Kumenosuke's sister Satsu, who has brought the ashes of their recently deceased father, but it is so dark she fails to recognize her brother. After parting from the sister, the lovers commit suicide.

The opening song of *maki* ii is an excellent example of how Chikamatsu uses short songs to introduce scenes, give the background of a situation, or suggest insights into the psychology of the characters. With subtlety and precision, the song maps out the structure of the entire scene in a few lines of verse.

Figure 17. Scene One from Love Suicides at the Women's Temple. *Kumenosuke is evicted from Mt. Kōya's sacred precincts. 1911 Asahi-za poster. Courtesy of Ōtani Library.*

fushi
Long ago, before they met, she was spotless as a sheet of blank paper,
chū
but then his secret visits grew thick as cardboard, only for others
u
perversely to rip across the grained paper, and her sleeves were all
fushi
too often soaked with tear paper. Many-tongued rumor quickly laid
u *u* *u*
her waste paper; weak before its storms as tissue paper, this tenderly
fushi *haru*
protected child is only seventeen, exquisite even in her name, Oume,

the plum blossom.[15]

A perceptive interpretation by Yūda of this song is that the word play on the various kinds of paper outlines the five-movement structure of the scene.[16] "Blank paper," he suggests, represents the parents' ignorance of Kumenosuke's plight in *maki* i (movement A). "Thick cardboard" refers to the layers of love letters between Oume and Kumenosuke (movement B).

"Torn paper" suggests the effects of the marriage arrangements, which separate the lovers (movement C). "Tear paper" refers to the mother's tears (movement D). "Waste paper before the storms as tissue paper" points to movement E, when the lovers are driven from the house. The outline is as follows:

A. Wedding night (blank paper) (to Keene, p. 144, "With these words she goes with Oume into the linen room.")
B. Secret meeting (thick cardboard) (to p. 147 "the song thrush sporting among the plum blossoms feels for people who interfere!")
C. Call off wedding (torn paper) (to p. 149 "The two grapple, now one, now the other on top.")
D. Mother's tearful intervention (paper wet with tears) (to p. 151 "feeling more dead than alive.")
E. Escape (waste paper blown in the storm)

Chikamatsu juxtaposes the festive wedding atmosphere and joyful spirits of the father against the sad, desperate Oume. The audience knows well, after watching *maki* i, the real situation, and would naturally be sympathetic towards the lonely girl, anxiously awaiting the arrival of her lover to save her from this dilemma. Initially, we know more than Oume, but gradually, as Oume learns what happened on the mountain, we are put in the same position as Oume and are caught up in the suspense of the impending crisis. The technique, again, is based on the *jo-ha-kyū* formula; the first two movements are introductory to movements C and D, which force the pair to choose flight and death, and bring the play to its climax. Movement C is violent, as Yojiemon and Sakuemon trade insults and then blows. At the moment of crisis, Oume's mother intervenes and pleads for sanity. At the same time, her words beg the lovers not to kill themselves. The mother's soliloquy is a *kudoki,* an impassioned display of love and grief, and is the emotional (and musical) climax of the entire play. Her heartfelt words, though, have a contrary effect; they thrust the couple toward destruction, because Oume's forced acceptance of Sakuemon as her husband gives the lovers no choice but death. The final movement is masterfully arranged. The scene in the dark is a *danmari* stylized dance pantomime and has plenty of suspense and excitement to titillate the audience. Again, the mother's love backfires and forces the pair out of the house into the jaws of death, wasted paper blown about in the raging storm.

Both *maki* i and ii end in darkness. In i, the time is daylight, but "heaven and earth are obscured by black clouds, and the world is shrouded in the darkness of endless night." At the end of ii, the time is near midnight, and the pair are thrust out into the night. The temporal sequence is exactly the same as in *Love Suicides at Sonezaki:* daylight–evening–deepest night–dawn, the action taking place in less than twenty-four hours. But this play has no sense of joy in death as in *Sonezaki.* Even at dawn, when the light of new life should shine on them after the turmoil of night, the sun is hidden by thick clouds.

This couple is not of the same social status as in most other love-suicide plays, because they are neither orphans nor social outcasts. Love drives them from society. In *maki* i, Kumenosuke gains our sympathy because of two factors: one, the hypocrisy of the monks, who condemn Kumenosuke's love for Oume, but who wallow in their lust for catamites; the other, the final dramatic scene, where Kumenosuke is violently cast down from the holy mountain by human and heavenly powers. All is lost for gentle love of a young maiden. In both *maki* i and ii, the pair appear as innocent, though impetuous, youths in stark contrast to their less sincere and less sensitive elders.

The whole play is almost circular, in that the action moves down the mountain to the town and back again for the final suicide; and the setting for the suicide is near the Women's Temple at the gate of the Mt. Kōya temple complex. For these lovers, however, death is not the glorified, romantic escape from a cruel world, as in *Sonezaki.* Rather, they are heartlessly forced from this world by all around them. The *michiyuki* is, accordingly, much darker and more solemn. It consists of five movements, outlined as follows:

A. *Makura*
   Introduction; imagery, "love lost in shadows," "crossing of Six Ways," "dark, misty moonlit night" (to Keene, p. 155 "like water to moisten the lips of the dying.")

B. Departure
   Pair pass familiar places; Oume and Kumenosuke express

sad feelings for her parents (to p. 155 "now we have reached Fudō Hill.")

C. *Kudoki*
Memories and Regrets: "Are we crossing the mountain road that leads to death?" "Scattered cherry blossom buds" (metaphors both for enlightenment and the death of the young couple). Oume regrets hurting her mother and brother Hana (to p. 156, "The wind blowing down India Mountain is his punishment on us; it cuts our flesh as we embrace.")

D. Buddhist prayer
"How pitiful these moments haunted by the terrors of hell." "By Snake Willow a thousand demons may assail us, but however tormented or afflicted, we shall not part," "undivided heart" (to p. 156 "sacred relics of his True Body.")

E. *Kiri*
Finale; "become Buddhas or wander by the Three Ways of Hell," prayer for salvation, "dawn moon is buried in the clouds of Five Hindrances" (to p. 157 "they have reached the Women's Temple."[17]

The scene is imaginatively in total darkness. To the lovers' regret, they cannot even see each other's faces.

The *michiyuki* is a song and dance scene and is always beautiful on stage, and certainly Chikamatsu's poetic language, carried by the music, was a delight to the audience. The opening *makura* introduces a particularly dark and Buddhistic atmosphere. The journey is through the world of "shadows," "misty moons," and the spiritual world of the "Six Ways." Perhaps it is easier to understand the significance of the *michiyuki* if we imagine it happening on two planes: first, the physical one of the two people actually traveling from Oume's home in Kamiya to the site of their death near the Women's Temple; and, second, the spiritual one of the souls wandering through the Buddhistic spheres of existence in the reincarnation cycle. During the spiritual journey through the nether world and the consequent purgatorial experience of suffering in hell, the pair gain the right to salvation after the death of their physical bodies. The third movement begins with the suggestion that the couple are crossing the mountain of death, and the dominant metaphor for

Figure 18. Suicide Scene from Love Suicides at the Women's Temple.
Sister (top) weeps when a stonecrop plant withers, because it is an omen that her brother Kumenosuke is dead.
1911 Asahi-za poster. Courtesy of Ōtani Library.

them is a falling cherry blossom, which is yet an unopen bud. In both movements B and C, the emphasis is on the misery of Oume who, by dying in this way, will hurt her mother and brother. Then, in movement D, the first line explicitly describes this spiritual journey "haunted by the terrors [along the road to] hell." But after this movement, the lovers cease looking backward, and turn to the future, where they envision themselves as one with an "undivided heart." This transition at the beginning of movement D from despair to hope is only afforded the pair after the spiritual journey through the terrifying nether world. In this play, however, Chikamatsu does not give the couple a happy, bright vision of the future, but instead ends the *michiyuki* in veiled darkness. The question of whether these two will become Buddhas or be condemned to wander in hell remains unresolved.

The scene after the *michiyuki* continues in darkness as the pair encounter Kumenosuke's sister Satsu, who does not recognize her own brother. Then the actual death of Oume is interrupted by cries of "murder," which tarnish somewhat the noble intentions of the lovers. Perhaps there is this absence of a bright vision of a happy future because the setting is in a Shingon Buddhism context, which does not place emphasis on heavenly paradise, as in Pure Land Buddhism. In any case, we are left with only the bleak image of two bodies returning to dust. Though this play is similar in many ways to *Love Suicides at Sonezaki,* it varies considerably in that death is no longer easy, no longer a blissful escape to paradise.

LOVE SUICIDES AT AMIJIMA

In the later *Love Suicides at Amijima* (1721), considered by many to be his masterpiece, Chikamatsu confronts most directly the question of what it means to die in love suicide. Although the play is a one-act *sewamono* in relation to the five-act *jidaimono,* it is actually longer than, or about the same length as, the three history plays (excluding *Coxinga*) discussed in the previous chapter. Yet it maintains a remarkable unity of tone throughout, and in that respect is similar to the short *Love Suicides at Sonezaki.* Chikamatsu himself appears to have considered *Amijima* classic, because in his last *sewamono, Love Suicides on the Eve of the Kōshin Festival,* he lists it with the traditional Japanese classics, *Essays in Idleness* and *Tales of the Heike.*

Love Suicides at Amijima begins in the pleasure quarter, where we meet a despondent courtesan, Koharu. Jihei comes looking for Koharu and overhears her talking with a samurai guest, who is really Jihei's older brother in disguise. The brother, Magoemon, comes to dissuade Koharu from committing suicide with Jihei. She asks him to become her patron for a year to keep Jihei from her. Her betrayal angers Jihei, who stabs a knife through the paper door, but fails to reach Koharu. Later, Jihei is ashamed when he realizes the man is his brother. Moreover, he is acutely embarrassed to have been beguiled completely by Koharu, and vows to forget her.

Maki ii is set at Jihei's home, where we meet his wife Osan, who is also his cousin, and their two small children. Magoemon and Osan's mother arrive with news that Koharu is to be ransomed; they are afraid Jihei is the buyer. Jihei pleads ignorance and vows never to see Koharu again. In the final scene of *maki* ii, Osan commands the stage. She is at first angry at Jihei for still loving Koharu, and then is remorseful at the thought of Koharu dying alone, due to her own secret letter requesting she leave Jihei to let him live. She encourages Jihei to go and save her. At this point, the stern father-in-law Gozaemon arrives and forcibly takes Osan away. In the final *maki,* Jihei and Koharu commit suicide.

Chikamatsu uses various structural devices to achieve the effects that have kept the play alive both on the stage and in the study for two and a half centuries. One technique is to organize the play within a framework of oppositions. As has been pointed out by commentators, the title of the play contains elements of a paradox.[18] The "*ten no ami*" ("heaven's net") refers, first of all, to a phrase in Chapter 73 of the famous Taoist treatise, *Tao te ching:* "Heaven's net is wide; coarse are the meshes, yet nothing slips through." This suggests an idea of justice, in which any deeds on this earth, good or bad, will return to the doer. It is a concept of retribution much like the straightforward Old Testament law of an "eye for an eye." But the title also refers to another net, the saving net in the vow of Amida Nyorai to help any and all who call Amida's name. If we continue the Biblical comparison, then this other half of the allusion is similar to the New Testament idea of mercy. To put the paradox simply, the two lovers, Jihei and Koharu, are simultaneously doomed and saved by their act of passion that causes so much suffering for those around them. In this play, Chikamatsu faces most directly the crux of this paradox.

The playwright reinforces this opposition by the arrangement of the settings of the three *maki:* i pleasure quarter–ii home–iii pleasure quarter–Amijima. The quarter is a world of desire and passion; the home a sphere of reason and responsibility. In each Jihei has a woman: Koharu in the quarter and Osan in his

Figure 19. Ukiyoe Actor Print (Toyokuni, 1849) of Love Suicides at Amijima. *First scene in the pleasure quarter. Jihei (Ichimura Uzaemon) and Koharu (Bando Shūka). Courtesy of Waseda Theater Museum.*

home. Until the beginning of the play he has been trying to live in both worlds. Shakespeare uses a similar structural technique in his "love-suicide" play, *Antony and Cleopatra,* where the lush world of the Nile is ruled by desire and erotic passion, and the world of Rome, by reason and military self-control. In Rome, Antony is a brave and commanding general, but in Egypt he is a lover, caring not for discipline and responsibility. For this pair, too, death is the final refuge.

Ordering a play into opposing worlds is a technique Chikamatsu uses often. In *Courier for Hell* (1711), the pattern is reversed. The pleasure quarter comes in the middle between Chūbei's two different home settings. The power of law (in the outside world) over the sphere of private desires (quarter) is projected clearly in the arrangement. This pattern, home–quarter–home, is also common in Kabuki plays of the time. The ordering of the settings usually reveals much of the play's meaning, because each *maki* gains further significance in juxtaposition. A good example is in the adultery play *Drum of the Waves of Horikawa* (1706). Chikamatsu in *maki* i suggests, through symbolism and action, that the scene is a dream (or nightmare) world of intoxication and illusion. Human desires, particularly sexual, find free release, and lead to the crime of adultery. *Maki* ii contrasts sharply with the previous scene, in that it is in the daytime, waking, and sober world of samurai law, where the crime of passion meets the sentence of death. The final *maki,* that of the vendetta against the male adulterer, significantly occurs during the Gion festival, a setting of daytime intoxication and freedom, when the everyday world is turned upside down. In this arrangement, the publicly sanctioned violent desires of the revengers are juxtaposed against the sexual desires of *maki* i, and the language used at the point of the murder is full of animal metaphors, as each attacker relishes the chance to thrust his or her sword into the drum teacher, "who falls on all fours like a dog."[19] Although Chikamatsu makes no overt comments on the action of the revengers, critical perspective is provided through juxtaposition. The twentieth century would find the comparison of sexual and violent desire interesting.

Since Chikamatsu consistently uses such techniques, we can be confident in our interpretation of its intent. In *Amijima,* the juxtaposition between home and pleasure quarter underlies what the Japanese call the opposition of *giri* and *ninjō,* translated variously as "duty" and "human feelings." "Reason" and "desire," however, seem more appropriate and meaningful, at least in the context of Chikamatsu criticism. Simply put, reason is what the mind (and society) determines is correct behavior, and is what makes civilized man different from animals or the uncivilized. Desires, on the other hand, are natural instincts, including love. Although what Chikamatsu's society considered rational may be different from what the ancient Greeks, Elizabethans, or contemporary Westerners did or do, this difference matters less than the fundamental opposition of these two concepts, which is a part of all cultures. For Jihei, and the other tragic figures who populate Chikamatus's love-suicide plays, there is no earthly solution to this dilemma, because society viewed it in absolute terms; equivocation was not a rational choice. In this play, although the choice is certainly death, Chikamatsu makes the consequences extremely painful and unpleasant. The lovers are never allowed to turn their backs on this world in order to look only to a bright future in the next life.

This play is different in that the central crisis is not whether the pair will commit suicide, only when it will happen. Before the opening lines, the characters Jihei and Koharu already have exchanged letters, stating their intention to die together. The song that introduces Jihei makes this clear.

> They swore in the last letters they exchanged that if only they could meet, that day would be their last. Night after night Jihei, ready for death, trudges to the Quarter, distractedly, as though his soul had left a body consumed by the fires of love.[20]

Therefore, dramatic tension lies less in the build-up to the choice of death, and more in suspenseful waiting for the chance to die. From this the play derives its power; the entire drama stands precariously on the threshold of death. The incident in *maki* i, when Jihei violently kicks Koharu in anger at her seeming betrayal, builds suspense—is Koharu serious and is Jihei really through with her? Although Jihei and Koharu do meet in *maki* i, they do not meet alone until they set out on the *michiyuki* journey to death. This is effective drama, because we, as well as the characters in the play, constantly confront the specter of death, waiting for completion of the known fact.

Yūda spent a long time analyzing and annotating this play. The fruits of his work have been collected into a book published after his death, *Zenkō "Shinjū ten no Amijima"* (A complete annotation of *Love Suicides at Amijima*).[21] He minutely analyzed the musical, dramatic, and textual aspects of the play, originally publishing this research in journal articles beginning in 1962 until his death in 1973. His ideas about musical structure seem to have evolved as his work progressed, since some ideas found in this book differ from those in his edited collection of Chikamatsu plays published posthumously. One problem is that Chikamatsu's plays written for Takemoto Gidayū have exact, predictable structures. The pattern of primary units (*ji* to *fushi*) is regular, and variations are usually significant, coming at the dramatic climaxes of scenes. After Gidayū's death in 1714, however, a change is noticeable, and its significance is puzzling. Was it a conscious choice of the chanters and playwright? And if so, what were the intended dramatic effects? A thorough examination of this trend is beyond the scope of this book, but in discussing *Amijima* and some other late plays, I will offer a possible perspective on this development.

The particular change to which I refer is in the makeup of a primary unit (see glossaries). The pattern is that, when a primary unit comes to a conclusion with a *fushi* musical cadence, if the primary unit is complete, *ji* or *ji iro* follows, and thus a new primary unit commences. The earlier examples, which I drew upon to explain the rhetoric of the words and music, were complicated primary units; they usually divded into two or more parts with *suete* or *fushi* notation followed by *kotoba* or a notation for pitch (*chū, u, haru*). In many late *sewamono* I have examined, there are numerous examples of *fushi* cadences followed by a notation for pitch or *kotoba,* where, in an earlier play from Gidayū's era, there would have been a *ji* or *ji iro*, and the start of a new primary unit. This tendency is pronounced in Chikamatsu's last three *sewamono* plays, *Love Suicides at Amijima, Onna goroshi abura jigoku* (Woman killer and the hell of oil, 1721), and *Love Suicides on the Eve of the Kōshin Festival. Maki* i of *Amijima* has, if we follow the *ji-fushi* notation, only six primary units, compared with seventeen for the much shorter first *maki* of *Love Suicides at Sonezaki. Maki* ii has but ten units, compared to twenty-two in *Love Suicides at the Women's Temple,* which is almost exactly the same length. Obviously, the units are much longer and more complex than the earlier ones. What different theatrical effect would these changes produce?

To consider this question, we must review again what happens during one of these units and what the intended effect is on the audience. The unit is based on the musical *jo-ha-kyū* pattern, which in this case means a slow and deliberate beginning, followed by a quickening of the pace and tightening of the tension to a point of climax, and concluded by a short relaxing cadence (*fushi*). Within the primary unit, the performer seeks to bring the audience to a particular emotional peak, and, after that is achieved, he allows himself and the audience to relax in preparation for another build-up. By not following a *fushi* (in this case incomplete) cadence with a *ji* line, and by not giving a slow beginning to the new unit, the performer stretches the tension, giving the audience less time to relax and step back from the illusion of the stage. This is crucial in a theater of puppets, where the edge between illusion and reality must be straddled by the audience, constantly crossing back and forth. In the puppet theater, when one stops and notices the wooden puppets and manipulators, the spell is broken—though it is in the slender margin between illusion and reality that the wonderful effect of Bunraku theater lies. Could it be chanters were trying to increase dramatic intensity by stretching the concentrative powers of the audience, thereby not allowing them time to step back from the illusion of the stage?

Since the pattern is a departure from the traditional norm, it is obvious they were experimenting with stage effects. In the tightly knit drama *Love Suicides at Amijima,* the lovers are only waiting for the chance to meet and to kill themselves. Perhaps the chanters sought to increase even further the tension in this very solemn play. Let us examine in detail the structure of *maki* ii. The setting is Jihei's home, where he lives with his wife Osan and their two children. Yūda divides the *maki* into four movements instead of five.[22]

A comparison of this outline with any other shows clearly how different the musical structure is. The pattern is for two short units (1,2) to be followed by a very long unit (3a–3d), or one single unit (4) followed by a series of long ones (5a–5c and 6a–6f). The effect on the audience would be subtle. Since the musical structure and conventions were clearly delineated, connoisseurs would have definite expectations as they watched and listened. A particular expectation (maybe not a conscious one) is that, after a full *fushi* cadence, a slow deliberate line would introduce a new passage. During this transition, the audience is offered a brief respite of relaxation, a moment of quiet between storms. If the chanter continues quickly after an incomplete cadence with speech or continues the pace with just a change in pitch, then this expected moment of relaxation is swallowed up in the continuing flow of the drama. Another way to divide this *maki* is in two parts, *kuchi* (A and B) and *kiri* (C and D). From Chart 10, we can see that unit 3 is the climax of the *kuchi,* and that unit 6 is the climax of the *kiri.* In unit 6, the performers must hold the complete attention of the audience for an exhaustingly long time at a level of extreme intensity, because there is a series of (incomplete) cadence endings and no beginnings (that is, *ji* or *ji iro*). If successful, it is a tremendously moving scene and the tragic underpinning of the entire play; if

CHART 10

Musical Structure of *Love Suicides at Amijima, Maki* ii[a]

A. Introduction of Jihei's home; comic scene
- 1. Explanation of stage setting in paper merchant's home (to p. 403, "customers come thick as raindrops.")
- 2. Osan runs affairs of home and shop (to p. 403, "She runs impatiently to the shop.")
- 3. a. The comic fool Sangorō (*fushi* p. 404 "the others can only smile despite themselves.")

B. Visit of Magoemon and mother; Jihei vows to forget Koharu
- 3. b. (*kotoba*) Entrance of Magoemon and mother (*fushi* p. 405 "Osan serve the tea!")
- c. (*kotoba*) Jihei is accused of ransoming Koharu (*suete* p. 406 "She collapses in tears of resentment.")
- d. (*haru*) Jihei writes out promise to leave Koharu; Magoemon and mother leave (*fushi* p. 407 "She leaves, her heart innocent as Buddha's.")

C. Osan's outrage at Jihei, and confession of sending letter to Koharu
- 4. Jihei weeps in *kotatsu* (p. 407 "deep enough to bear him afloat.")
- 5. a. Osan's outburst of anger and desperation (*fushi* p. 408 "throws herself over him, moaning in supplication.")
- b. (*haru u*) Jihei's honor vis-à-vis Tahei (*suete* p. 408 "He collapses with weeping.")
- c. (*haru*) Osan's confession of sending letter to Koharu (*fushi* p. 409 "She melts in tears.")
- 6. a. Osan gives Jihei money to ransom Koharu (*fushi* p. 409 "this money he never put there.")
- b. (*kotoba*) Osan gives up all possessions for Jihei (*suete* p. 410 "and puts her love in besides.")
- c. (*kotoba*) Osan bares her soul to aid Jihei (*chū fushi* p. 410 "and he is silently weeping.")
- d. (*kotoba*) What shall Osan do? Be his cook? (*suete* p. 410 "She falls to the floor with a cry of woe.")
- e. (*kotoba*) Jihei begs forgiveness (*fushi* p. 411 "I don't regret it in the least.")
- f. (*ji iro chū*) Osan with no regrets hurries him out the door (*fushi* p. 411 "tonight it will be stained with Koharu's blood.")

D. Gozaemon takes Osan away from Jihei
- 7. Entrance of Gozaemon (*chū fushi* p. 411 "Husband and wife are upset and confused.")
- 8. Gozaemon's anger (p. 412 "stunned into silence.")
- 9. a. Jihei begs forgiveness (*fushi* p. 412 "his face is pressed to the matting in contrition.")
- b. (*kotoba*) Gozaemon finds all family possessions gone (*fushi* p. 413 "ready to sink into the fire of humiliation.")
- 10. a. Gozaemon demands a divorce (*suete* p. 413 "raises her voice in tears.")
- b. (*chū u*) Osan is dragged away from her family.

Note: [a]Capital letters mark movements; arabic numerals, primary units; and small letters, subdivisions within the units. Ending and beginning notations are also given when the primary unit contains subsections. Page references are to Keene, *Major Plays of Chikamatsu.*

unsuccessful it becomes exaggerated melodrama. The play's popularity attests to its success on the stage.

Why this departure from tradition? Perhaps it was due to the influence of Gidayū's successor Masatayū, who appears to have had a delicate, rather than powerful, voice. He was interested in the finer and subtler aspects of character. The treatise containing his ideas, *Ongyoku kudensho* (Oral teachings on music), shows this preference.[23] It consists of interpretations of scenes and characters and admonitions for chanters to consult the playwright to gain a precise understanding of the text. He uses the phrase, "*jō o fukaku kataru*" (chant deeply the essence of emotion), as the cornerstone of his art. This attitude perhaps can help explain the performer's intention in unit 6. Although the play is about the love suicide of Jihei and Koharu, the climax focuses on the character Osan, the neglected point in the love triangle. Osan sacrifices part of herself for her husband and Koharu in each section of the unit: first, her money; next, her and her children's possessions; then, her soul; and, finally, her dignity. Each is offered freely, until section 6d, when she collapses under the weight of grief. The absolute nature of her devotion and consequent grief adds poignancy to the final suicide. In unit 6, emotions are stretched to the limits of pity and sympathy. The section in 6b, when Osan takes all the family's few remaining possessions from the chests, is an arialike passage expressing her feelings. Yūda makes the adroit observation that, as the passage progresses and her feelings of grief grow keener, the lines become longer,[24] certainly an example of the play's intricacy and depth. The playwright and performers go to great lengths to direct the audience's emotions. Unit 6 is too long to quote in full, but I will give about half of it, from 6b to 6d, and include the musical notation to show the rhetorical methods employed. (See glossaries for explanation of notation.)

kotoba

6b (Osan) I'll tell you later where this money came from. I've scraped it together to pay the bill for Iwakuni paper that falls due the day after tomorrow. We'll have to ask Magoemon to help us keep the business from betraying its insolvency. But Koharu comes first. The packet contains 400 *momme*. That leaves 350 *momme* to raise.

ji iro u *haru*
(Narrator) She unlocks a large drawer. From the wardrobe lightly fly kite-colored Hachijō silks; a Kyoto crepe kimono lined in pale brown, insubstantial as her husband's life which flickers today and may vanish tomorrow; a padded kimono of Osue's, a flaming scarlet inside
chū *u*
and out—Osan flushes with pain to part with it; Kantarō's sleeveless, unlined jacket—if she pawns this, he'll be cold this winter. Next
haru
comes a garment of Gunnai silk lined in pale blue and never worn, and then her best formal costume—heavy black silk dyed with her family crest, an ivy leaf in a ring. They say that those joined by marriage ties can even go naked at home, though outside the house
u
clothes make the man: she snatches up even her husband's finery, a silken cloak, making fifteen articles in all.

(Osan) The very least the pawnshop can offer is 350 *momme* in New Silver.

u
(Narrator) Her face glows as though she already held the money she
suete
needs; she hides in the one bundle her husband's shame and her own
noru chū *haru*
obligation, and puts her love in besides.

kotoba
6c (Osan) It doesn't matter if the children and I have nothing to wear. My husband's reputation concerns me more. Ransom Koharu. Save her. Assert your honor before Tahei.

ji chū *haru* *chū fushi*
(Narrator) But Jihei's eyes remain downcast all the while, and he is
haru
silently weeping.

kotoba
6d (Jihei) Yes, I can pay the earnest money and keep her out of Tahei's hands. But once I've redeemed her, I'll either have to maintain her in a separate establishment or bring her here. Then what will become of you?

iro
(Narrator) Osan is at a loss to answer.

kotoba
(Osan) Yes, what shall I do? Shall I become your children's nurse
ji u *ue*
or the cook? Or perhaps the retired mistress of the house?
suete *chū*
(Narrator) She falls to the floor with a cry of woe.[25]

Section 6b begins in straightforward speech style, *kotoba,* as Osan matter-of-factly discusses finances and Koharu's ransom. The passage, beginning with *ji iro u,* describes Osan as she gathers together the family's last possessions to take to the pawnshop for money to buy Koharu's contract. The emotional and lyrical tone contrasts with the previous lines. The passage is a conventional "catalogue of materials" song, and in this case the materials represent the crumbling family's fate; without sufficient clothes, her children will freeze during the next winter. At the end of 6b, Osan is a strong, resilient woman bearing up to her fate, and the language is forceful; however, the *suete* notation implies harsh music and suggests that under the noble surface writhes a tortured heart. The actor or puppet operator must put on a virtuoso performance to portray both sides of Osan—her tough determination, her underlying agony at assisting the destruction of her family. She is able to maintain a façade until she encounters the dilemma, in 6d, of what to do if Koharu actually comes to live in their home. Again, the pattern of beginning with cool speech and switching to hot song is the rhetorical technique. A cool voice continues until she considers seriously her role in the new household, and then the voice rises in song. *Ue* is the highest pitch, and, since it is followed by the harsh *suete,* we can imagine how the performers were attempting to bring the audience into sympathy with Osan.

In the two *sewamono* plays after *Amijima,* similar techniques of stretching the primary units are used, but apparently with less success. In *Woman Killer and the Hell of Oil,* the climactic scenes in *maki* i of Yohei's disinheritance, and in *maki* ii of the murder, both have very long units of the sort just analyzed. In Chikamatsu's last *sewamono, Love Suicides on the Eve of the Kōshin Festival,* this tendency is further extended. *Maki* i and iii are quite different from the earlier regular pattern, although *maki* ii follows traditional form. *Maki* iii is especially interesting, because there are only five primary units which correspond exactly to the five-movement divisions. The play is a particularly somber tale of a married couple—the wife is pregnant—who are driven to suicide by the mother-in-law, and not very exciting in terms of dramatic action. One could surmise that the playwright and performers tried to heighten the tension by this stretching technique. The plays of Chikamatsu's contemporary, Ki no Kaion, however, do not show this tendency. Just a few weeks before Chikamatsu's *Kōshin Festival,* Kaion's *Shinjū futatsu hara-obi* (Love suicides and the double-brocade obi, 1722), which was based on the same suicide incident, was staged. It appears to have been more popular than the later Chikamatsu version. Kaion's play is a bit more exciting and a better-written play, according to Donald Keene.[26] Perhaps for these reasons and the fact of its earlier debut, Kaion's play was more popular. Neither of these last two Chikamatsu *sewamono* were popular in the playwright's lifetime. Later Tokugawa audiences, however, did prefer Chikamatsu's *Kōshin Festival* to Kaion's *Double-Brocade Obi.*[27] Aside from the usual reasons given for their lack of success on the stage, perhaps the variation away from the traditional musical pattern failed to strike a responsive note in the audiences of that time. In *Amijima,* which is an extremely tightly unified play, the technique worked, but in the two later, less intense *sewamono* it was less successful.

The crucial *michiyuki* journey to death, contained in *maki* iii of *Love Suicides at Amijima,* follows the typical pattern found in love-suicide plays—a symbolic movement down to hell, up to death at dawn, and a culminating vision of happiness in paradise. As the pair are escaping from the quarter to seek their death, the text describes them forging forward, where "an inch ahead are the tortures of hell."[28] The following is an outline of the *michiyuki's* structure.

A. *Makura*
 Objective perspective on lovers' plight; those who travel the quarter meet this end (Sutra of Cause and Effect). Their suicide will be news for gossip, plays and books. Jihei is led on by the spirit of death, but is drawn back to his home (to Keene, p. 419 "it is hard to keep walking on.")

B. Path of darkness
 Moon is full but they cannot see because of their darkened hearts. Jihei must again reject his family. Lovers recall their three years of love and consequent suffering. "Is this already the road of hell?" (p. 420 "threatens to immerse even the Horikawa bridges.")

C. Suffering in nether world
 Jihei passes near his home but turns away again from family. Pair are frightened by demons (Temma) (p. 420 "for us about to depart this world.")

D. Faith in Buddha
 Jihei imagines death by sword and crossing river of death. Koharu expresses her faith in the Lotus Sutra. Vision of future life together; begin upward movement. Pair cross Kyō (Sutra) Bridge and arrive at "opposite shore" (Nirvana), and cross Onari (become Buddha) Bridge (p. 421 "They cross Onari Bridge.")

E. Arrival at Amijima
 Temple bell signals coming dawn.

During this *michiyuki,* entitled, "The Farewell Journey of Many Bridges," twelve bridges are mentioned, though the pair cross only six. Each bridge marks an entrance into a different world. The movement continues downward until they cross Temma Bridge. When the Kyō (Sutra) Bridge is crossed, they enter the merciful world of Amida. At the beginning of the journey reference is made to the Cause and Effect (Karma) Sutra, which Chikamatsu is perhaps comparing to the Taoist "heaven's net" of the play's title, through which nothing escapes. Jihei and Koharu are doomed by their actions. Yet, in the last part of the *michiyuki,* faith in the Lotus Sutra leads the pair into Amida's saving net. Their suicides are opposite to the path of Buddhism, yet their journey leads to Amida's paradise.

The area of Tenjin Bridge is Jihei's neighborhood. "Tenjin" also suggests the legend of Sugawara no Michizane and the famous three trees: plum, cherry, and pine, which Chikamatsu uses here to symbolize fidelity and love. The trees had all responded to their master's exile by either flying to his side or withering at his former home. These symbols contrast sharply

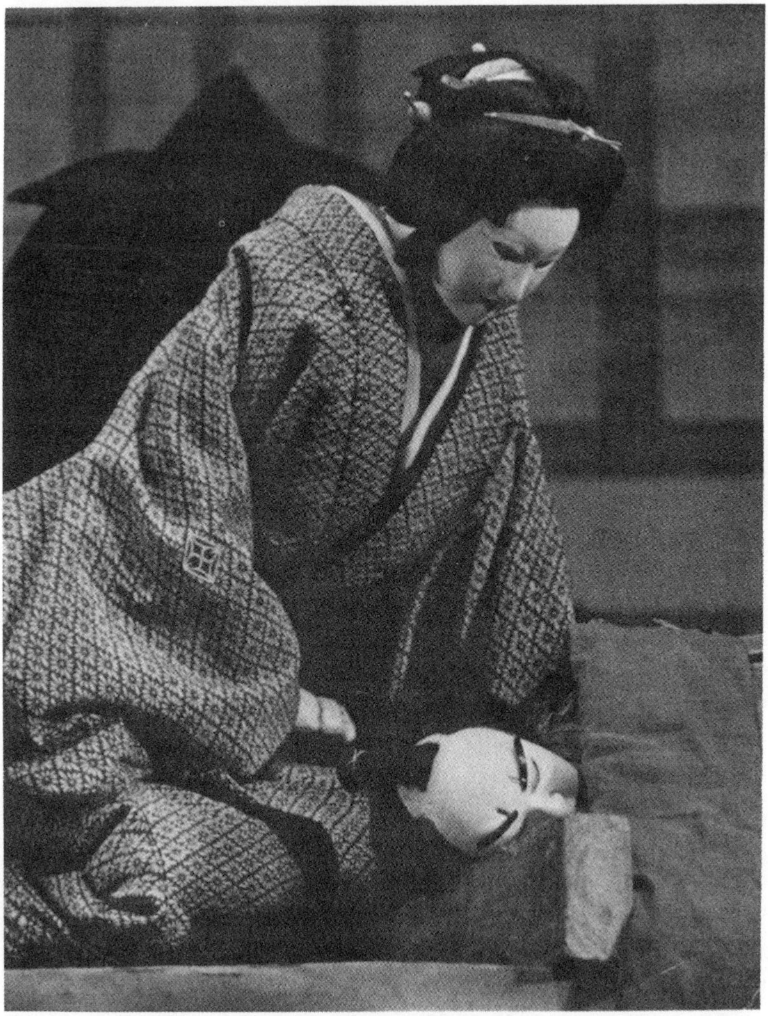

Figure 20. Contemporary Bunraku Performance of Love Suicides at Amijima. *Scene from* maki *ii, where Osan is angry at Jihei. Courtesy of Ikoma Takami.*

Figure 21. Utamaro print of Michiyuki *from* Love Suicides at Amijima. *Courtesy of Tokyo National Museum.*

with the negligent Jihei, who abandons his home and family. As they continue, the lovers' sadness gradually increases, and they imagine themselves along the road of hell. In movement C, Jihei is forced again to reject his family when they pass a few blocks from his home and immediately are frightened by demons. Movement D begins with a vision of death by sword and the consequent crossing of the River of Death. At this moment of suffering, however, Koharu expresses her faith in a happy future together, a view founded on her devotion to the Lotus Sutra and the mercy of Buddha. With her declaration of faith the journey turns upward, and the pair cross the Sutra Bridge and the Bridge (Onari) that symbolically makes them Buddhas. The place they choose to die is the "other shore" (nirvana), Amida's paradise (Amijima). Suffering leads to this resting place. The symbolism, from this life to the other world, brings to mind the icon at Enryakuji Temple on Mt. Hiei; the journey through the dark ravine leads to the bright world of Buddhist enlightenment.

This *michiyuki* is close in its design to what Yokoyama calls "retrospective summation." The dramatic actions of the play are the incessant attempts by those around Jihei to dissuade him from committing suicide with Koharu and thereby ruining his family. Magoemon's deception in *maki* i, Magoemon and Osan's mother's pleas and Gozaemon's demands in *maki* ii, and Magoemon's desperate attempt in *maki* iii of bringing the children along to extort pity from Jihei, comprise all the dramatic action. And this is the theme of the *michiyuki;* Jihei, once again, must face the consquences of his action and reject his moral and rational responsibilities to die for love.

The play's power emanates from its theme of the unrelenting confrontation with death. Having made the choice before the curtain opens, the couple's every thought is on this course. Chikamatsu, however, makes the journey excruciatingly difficult by blocking their every move to meet and die for love and by constantly reminding them of the consequences of their act. And that is why this work, in particular among the love-suicide plays, is so moving; it is a journey through a purgatorial realm where the lovers are forced to suffer the consequences of passion. For that reason it is a harsh play, but in this way the paradox of opposition implicit in the title "ten no Amijima" is resolved. The drama travels between retribution and mercy, and, at the end of this journey, the pair choose to die: one on a mountain, the other by a river. This is the only example among Chikamatsu's or Kaion's love-suicide plays in which lovers voluntarily choose to die physically separated. The usual case is for the pair to achieve a total physical (and sexual) union of their bodies at the moment of death, symbolized by a tree trunk growing from two different sets of roots. Shinoda Masahiro in his movie *Double Suicide* (Shinjū ten no Amijima, 1969), based on this work, has Jihei and Koharu experience sex in a graveyard before dying. This is a plausible variation within the love-suicide tradition, but it is a misreading of *Love Suicides at Amijima.* At the end of Chikamatsu's play two people, a priest and a nun, choose to die for a noble love which transcends the physical.

Chikamatsu's depiction of love suicide in *Amijima* is quite different from that in *Sonezaki* where dying for love was romantic, untarnished beauty. Tokubei and Ohatsu innocently saw death as a fortuitous escape from this world of suffering to a paradise of eternal joy. In *Love Suicides at the Women's Temple,* Chikamatsu still has sympathy for the plight of the unfortunate young couple, but he constructs a difficult road to death and clouds the romantic vision by inviting doubts on the possibility of reaching paradise in the afterlife. In *Amijima,* Chikamatsu shifts his focus from the lovers about to die to those left behind to go on living with painful memories. In *Amijima,* Osan's role in the last half of *maki* ii contains the essence of the play. She is a cousin and childhood friend, who has grown up in the same household with Jihei and can hate neither him nor the woman he now loves. The depiction of her suffering, as the one abandoned, is the climax of the play. Chikamatsu emphasizes the importance of life by having the pair, who are bent on suicide, suffer continually along the road to death. They are never allowed to forget the fact that their love is causing anguish for the innocent left behind. In the *michiyuki,* we glimpse again the trials of their past lives during the extended moment before death; this vision is of a tortuous journey through the nether world which leads eventually to the merciful, guiding light of Amida. They are forgiven. However, Chikamatsu does not have two die together as one, implying they will be unified forever;

each dies alone, a priest and a nun, with no ties to anyone anywhere. After the darkest moment of night, a new dawn does break for them and there is no doubt these two will reach salvation, but death is not a happy escape to paradise; rather, the specter of death affirms the preciousness of life.

Illusion for romance . . . is an order of existence that is best called alienation. Most romances end happily, with a return to the state of identity, and begin with a departure from it. Even in the most realistic stories there is usually some trace of a plunge downward at the beginning and a bounce upward at the end. This means that most romances exhibit a cyclical movement of descent into a night world and a return to the idyllic world, or to some symbol of it like marriage.

Northrop Frye, The Secular Scripture

6 · *Circles of Felicity*

YOSAKU FROM TAMBA

Seven of Chikamatsu's contemporary-life dramas have happy endings.[1] Most critics, Japanese and other, however, have considered these happy endings anything but felicitous. No one has felt comfortable with these "problem" plays with their seemingly contrived and unsatisfactory finales. Unlike most *sewamono* which depict a current happening such as a love suicide or a murder, these dramas are based on factual incidents of a generation or more ago drawn from contemporary ballads or works of fiction. Another distinguishing feature of these dramas is that the male protagonist, in most instances, is of a higher social status from those in love-suicide plays.[2] These "comic" plays are generally about the same length as the late love-suicide dramas but shorter than the history plays of Chikamatsu's mature period. Most have a three-*maki* form similar to the love-suicide genre, but have happy endings in common with the history plays. They do not fit neatly into any category.

Since these works contain elements of love suicide, adultery, (Kabuki) *yatsushi,* and crime plays, a thorough study of this group would demand more space than this one chapter. I have chosen two to discuss, *Tamba Yosaku matsuyo no Komuro-bushi* (Yosaku from Tamba, 1708) and *Yamazaki Yojibei nebiki no kadomatsu* (The uprooted pine, 1718), because they are in English translation and because they are fascinating dramas.

The *yatsushi* (disguise) convention, fundamental in these two plays, is similar to the tragi-comedy of Elizabethan theater or the Western genre of romance, in which the main character or characters fall on hard times and are brought to the edge of disaster, but are, in the end, always restored to their rightful high position in society. In Western romance, in particular, the similarity is striking; protagonists are invariably of the upper class and experience a loss of identity and social status until the conclusion of the tale. Although there are many differences between these unrelated traditions, many aspects of Western romance are evident as well in Chikamatsu's plays. In *The Secular Scripture: A Study of the Structure of Romance,*[3] Northrop Frye describes the fundamental movement of romance as a descent into a demonic world followed by an ascent to an idyllic paradise—a pattern found also in Jōruri. His analysis of the theme of descent in romance as a "coming down from a state of innocence or freedom into a 'prison house,' as Wordsworth calls it, of corruption or confusion"[4] describes accurately the situation of the male figure in Chikamatsu's plays.

Yosaku from Tamba begins in a daimyo's palace, where a wedding party is about to depart for Edo. The lord's ten-year-old daughter is to be married, but the little girl has no intention of going. Sankichi "the wild one," an eleven-year-old packhorse driver, is called in to play a game (*fushigoto* song) about traveling to Edo. After his entertaining show, the girl is cheered and ready to depart. Shigenoi, the girl's nurse, rewards the boy and during the encounter realizes that he is her son who had been evicted from the palace many years earlier due to his father's (Yosaku's) profligacy. She wants to accept this little ruffian as her son, but is blocked by her duty to her lord; Sankichi's presence would shame the girl and ruin the wedding plans. The parting of this pair is the climax of *maki* i.

Maki ii is set along the Tōkaidō road to Edo. Here we meet Koman, a low-level prostitute who is in love with Yosaku, a former samurai reduced to being a packhorse driver. She is poor and has a father in debtor's prison. Yosaku reveals that he has lost a large amount of money and his horse at gambling. The villainous Hachizō demands the horse, and, to save Yosaku, Koman gives Hachizō her hard-earned savings, which had been intended for her destitute father's welfare. Yosaku later asks

Sankichi (whom he does not realize is his own son) to steal money for him. The boy tries, but is caught and roughly kicked unnecessarily by Hachizō. Sankichi refuses to accept the pardon which Shigenoi's pleadings get for him, and instead kills Hachizō because of the shame brought on by the beating. He is then led away to prison. When Yosaku returns and learns of the terrible happenings, he and Koman set out to end their lives as penitence for having killed the boy.

Maki iii begins with their journey to death, but, when they stop to die, samurai catch up with them, interrupting their plans. The young princess, upon learning the background of Yosaku and Sankichi, pardons father and son as well as Koman. Yosaku is restored to samurai status, and at the end of the play Yosaku and Koman lead a dance in which the tale is of the play, *Love Suicides at the Sunken Well,* performed the year before.

Yosaku from Tamba has remained alive on the Kabuki and Bunraku stages to the present day. After Chikamatsu's lifetime, however, the play was staged as part of another longer revision *Koi nyōbō somewake tazuna* (A cherished wife and colored reins, 1751), but many of the dramatic scenes of the original, including the one of the encounter of Shigenoi and her long-lost child Sankichi, were incorporated virtually unaltered. These scenes have been praised by modern critics from Tsubouchi Shōyō to Donald Keene. None, however, has liked either the fanciful ending, a *deus ex machina* happening in which the young princess saves Yosaku and his lover Koman from the jaws of death, or the final "Yosaku odori" dance, which everyone contends has nothing to do with the drama. Donald Keene, following the line of Japanese critics, decided not to include this dance in his English translation. Therefore, although many have praised parts of it, all have considered the work as a whole flawed.

Critics and editors have been uneasy with this play and often tried or wished to alter it. The editors of one of the earliest annotated editions of this work, in *Chikamatsu no kenkyū* (A study of Chikamatsu, 1900), mistakenly believed that the final dance was appended to the play during its revival to give a lively ending.[5] Accordingly, they did not provide annotation for this portion. Wakatsuki Yasuji says that he considered the ending unnatural, but the earlier part masterfully written.[6] In an

Iwanami edition of Chikamatsu's plays,[7] the editor criticizes the work, saying that, although there are many fine parts, overall it is not one of Chikamatsu's better dramas, and that the fanciful ending is like a children's fairy tale. He criticizes the final dance as well, saying that it is only a summary of the play *Love Suicides at the Sunken Well* and is like "bamboo grafted to a tree." Hirosue, in a more detailed critique, follows these earlier scholars in praising some parts and in condemning the happy ending because it destroys the unity of the tragedy.[8] He suggests that this play has two separate themes: Shigenoi's dilemma of being caught between love for her child and duty to her lord and Yosaku and Koman's love story. Hirosue blames Chikamatsu for failing to write a tragedy suiting Hirosue's own definition, and fails to approach the play on its own terms. To gain a perspective on the problem of this work, however, we need to discover what kind of play Chikamatsu was attempting to create. What dramatic conventions brought the drama to a happy conclusion?

Although the editors of *Chikamatsu no kenkyū* had their doubts about the final "Yosaku odori" dance, they felt the structure, beginning and ending with the princess, to be well ordered and symmetrical.[9] Hirosue suggests that one should compare it to Kabuki or Jōruri history plays,[10] although he himself does not follow this wise suggestion. The play's symmetry, Hirosue's idea of two themes, and his suggestion of comparing *Yosaku* to Kabuki and *jidaimono,* offer some directions on how to approach this drama. Let us first look at the play's sources.

Two scholars, Kuroki Kanzō and Suwa Haruo, have investigated the background of the "Yosaku" theme.[11] Though the factual origins of the character Yosaku are obscure, numerous songs, dances, and Kabuki plays bear his name. A poem from an early ballad to which Chikamatsu alludes in the *michiyuki* states: "Yosaku from Tamba was a horse driver, but now he's a sworded samurai from Edo." This song appears before and after the time of Chikamatsu's play in different collections of songs. Another song that Chikamatsu quotes in the *michiyuki* is about Koman: "When she thinks of you, Yosaku, / The shining sun clouds over– / Is that shower of tears / From Koman of Seki?"[12] Suwa suggests that these two stories, one about Yosaku, the other about Koman–originally separate tales–were joined together before Chikamatsu's time,[13] but he offers no evidence. The Kabuki repertoire contains numerous examples of Yosaku plays before Chikamatsu's version, but only one of these is extant today. In an earlier play, dating from 1677 (no longer extant), the actor Arashi San'emon is described as having played the role of Yosaku in the *yatsushi* manner.[14] All the other examples of Yosaku plays too are in the *yatsushi* pattern: Yosaku loses his high samurai status, falls into the low life of the pleasure quarter, and is eventually restored to his rightful position at the end. The extant play by Tominaga Heibei, *Tamba Yosaku tazuna obi* (Tamba Yosaku and a sash for reins, 1693),[15] a typical example of the *yatsushi* genre, is a fanciful drama that obviously had the most direct influence on Chikamatsu's work.

This Kabuki play is divided into three acts, with a variety of scenes. Shuzen (later Yosaku), the heir of his household, has a scheming stepmother whose insidious machinations are the source of many troubles. The stepmother tries, through various means, to destroy any heirs of Shuzen so that her own child can succeed to the household. When Shuzen's consort, Okiku, becomes pregnant, the stepmother has a *yamabushi* priest cast a curse to keep the child from being born. Okiku, evicted from the house by the stepmother, is wandering about the country at the beginning of Act I. Two servants accidently find her on their way to pray to Jizō for Shuzen's official wife, Oroku, so that she might conceive an heir. Okiku tells them her sad story, rips apart her protective charm, and collapses. From within her womb, the child's voice explains that a *yamabushi*'s curse is blocking his birth. In the next scene, the stepmother has Okiku killed, and from the woman's slashed neck the boy is supernaturally born. In the third scene, the stepmother has Shuzen disinherited, and he becomes a horse driver with the new name of Yosaku. He appears with the child, accompanied by the spirit of the deceased Okiku which has taken a temporary human form. Yosaku helps a stranger called Hannojo in his attempt to exact revenge upon his father's murderer. The man Hannojo seeks is the evil *yamabushi* who cast the curse on Okiku.

In Act II, Oroku, who has fled from the stepmother, meets Yosaku and the child. Hannojo conveniently forgets Yosaku's earlier favor and tries to wrest Oroku from Yosaku. Oroku helps the injured Yosaku to flee from Hannojo. The couple want to commit suicide, but, when a priest tells them of Hannojo's approach, they are unable to complete their plans. During the ensuing fight, Yosaku's young son manages to kill Hannojo. In the final Act III, the stepmother plots to make Yosaku commit suicide, but she is captured instead. The ending is a victorious dance and triumphant march into Edo. And, at that point, there is the famous line about Yosaku having been a horse driver, but now being a samurai in Edo.

This brief outline reveals some of the antecedents for Chikamatsu's characters and dramatic incidents. Kuroki suggests that Hannojo is Chikamatsu's Hachizō, and Okiku is Koman.[16] Oroku could be an antecedent for Shigenoi, and Okiku's child, who had such an unusual birth, is comparable to Sankichi, "the wild boy." In both plays the son kills his father's enemy. There is even a scene in the Kabuki where the pair Yosaku and Oroku contemplate suicide. This Kabuki play is obviously Chikamatsu's major source for *Yosaku from Tamba.*

Before examining Chikamatsu's work in relation to these antecedents, we should note what happened to the Yosaku legend after Chikamatsu. Kuroki explains that, although the later "Yosaku" Kabuki plays remain close to Chikamatsu's version, they are invariably expanded and changed into *oiesōdō yatsushi,* of the type before Chikamatsu's play.[17] In Jōruri as well, the tendency, as Toita Yasuji has pointed out,[18] is for the plays to follow the *yatsushi* pattern of a fall from a higher status at the beginning and a return to that status at the end. The version most popular on the Kabuki or Bunraku stages today, *Koi nyōbō somewake tazuna,* dates from 1751. It is a long vendetta play in the *jidaimono* style, although it is a revision of Chikamatsu's *sewamono* play. Chikamatsu's play is exceptional in its attempt to bring the material into the *sewamono* genre, however, it does not seem odd if it is seen as a Kabuki-style *sewamono.*

The structural characteristics of a *yatsushi* drama are similar to the history play: both begin and end in the same place after a journey through a lower world of trials and tribulations. The protagonist, in this case Yosaku, loses his samurai status and must endure many sufferings and indignities as a horse driver before returning to his rightful status. Within the *yatsushi* genre, the convention of a happy resolution is completely acceptable to Kabuki or Jōruri audiences. The problem, if one considers this play in those terms, is that Chikamatsu has mixed two quite different genres and not completely succeeded (at least from a reader's viewpoint) in his attempt to bring them together satisfactorily. The play is a conventional mélange of the *yatsushi* and love-suicide play. The two famous poems correspond to these two different genres. In the first, the horse driver who becomes a samurai again is from the *yatsushi* tradition; in the second, the lowly prostitute Koman who weeps for Yosaku, suggests the love-suicide genre.

The conventions, upon which these two plays rest, are not only different but are in opposition, because the love-suicide play demands factual realism, namely suicide, while the *yatsushi* play demands a restoration of the main characters to their former social status, whether the ending is realistic or not. At the heart of this conflict is the dilemma of the samurai Yosaku, whose duty on the one hand is to his lord only, but who must contend on the other with his love for Koman which demands fulfillment in death. Yosaku must stay alive for duty, and he must die for love and for his son whom he believes he has murdered. Chikamatsu further bolsters this opposition by having the women in Yosaku's life represent two extremes of duty: Shigenoi to a lord and Koman to love. Between these two paragons flounders a weak, frivolous Yosaku; even his name "Date (dandy) no Yosaku" reflects his personality. As in other plays, Chikamatsu juxtaposes themes from *maki* to *maki* to put this opposition in relief. *Maki* i, which takes place at the daimyo's residence, reaches its climax with Shigenoi's struggle to overcome her natural feelings of love for her long-lost child in order to remain dutiful to her lord. *Maki* ii, in contrast, is set along the Tōkaidō road, where the main figures are the raw and lusty characters of the working classes. The dilemma centers on Koman, who gives up all her savings to preserve Yosaku's honor, and, in so doing, condemns her father to die in a debtor's dungeon. Yosaku, on the other hand, inadvertently causes the death of his own son Sankichi. *Maki* iii, then, contains alternately both of these themes of samurai duty and love suicide; the *michiyuki* is like that of a love suicide, but the following scene focuses on the question of a samurai's duty. The final "Yosaku odori" dance returns the action to the love-suicide theme. Although this juggling of themes appears somewhat awkward from a critical point of view, it is interesting to see how the playwright was trying to reconcile the two contrasting themes and conventions.

In Chikamatsu's history plays, the drama returns to the place of origin after a journey through a dark world of suffering. The experience ideally will change the characters and audience, giving them new insights into the ways of mankind. Since Yosaku, too, returns to his origins as a samurai, we can expect that Yosaku will be somewhat altered by the end of the play.

The ideal samurai is a selfless warrior who devotes himself solely to his lord without thought to personal glory or disgrace. Such a code, however, denies natural human feelings. In *maki* i this dilemma is exemplified in the encounter of Shigenoi and Sankichi. Sankichi, a "natural" boy, cannot comprehend his mother, who speaks double talk—although she is his mother, she is now no longer his mother. When Shigenoi offers Sankichi money even though she no longer openly considers him to be her son, Sankichi refuses to take money from "strangers." Resolute in his attitude, he is a kind of noble savage who lives by a "natural" code of honor, straightforward and from the heart. The figure of Sankichi, acknowledged by those around him as a true samurai, gives the audience an enlightening perspective on the samurai class as a whole. Yosaku, in contrast to both these characters, appears as a dishonorable figure who has wronged them due to his profligacy. Before Chikamatsu's play begins, Yosaku has already lost his high samurai position due to his uncontrollable passions. Not until the final scene when he denies himself completely, including his desire to die, does he regain his former status.

The path leading to Yosaku's turnabout is traced in *maki* ii. Yosaku asks young Sankichi to help him steal some money, unaware that the boy is his son. During the attempt, Sankichi is caught, but is pardoned due to Shigenoi's tearful intervention. All appears well until Sankichi lashes out and kills the villainous Hachizō, who had earlier kicked and insulted him. This action cannot be pardoned, and the boy is condemned to die. Yosaku appears and is willing to take the blame, but his arrival is too late. This turn of events shocks Yosaku and Koman and drives them toward suicide by which they hope to atone for their crime of killing the innocent boy. The final lines of this *maki* describe the road they will take "Now they will travel the Six Ways with a relay of horses; they will cross the River of Three Roads; and the mountain ahead is not 'Earth Mountain in between' of the old song, but the mountain of death. They journey by through-horse to hell, oh world of dreams!"[19] The "Earth Mountain" song, which appeared at the end of *maki* i, symbolized the sadness and pain that lay between Shigenoi and her child Sankichi and between the opposing forces of love and duty. At the end of *maki* ii the power of the metaphor is intensified, for it no longer represents just sadness at parting, but the ultimate tragedy of death.

Chikamatsu makes it clear that the journey of the lovers to the site of their suicide is through hell. This pattern is the same as in the *michiyuki* of love-suicide plays. Yokoyama, in analyzing various love-suicide *michiyuki,* found this one structurally no different from the others.[20] Yūda divides it into five movements.[21]

A. Introduction (*Makura*)
Yosaku's song; dark, downward journey (to Keene, p. 124 "He'll surely stop at Yokkaichi.")
B. Yosaku and Koman sadly recall their past (p. 125 "We shall bury our fame in Sunken Fields.")
C. The sad destiny of their pledged love (p. 125 "pine groves of Ano when we pass.")
D. Vision of death (p. 125 "Her eyes look at him through tears—what was that?")
E. Prayers and dawning vision of afterworld "Inn of Hundred Delights" (p. 126 "they reach the Pine of a Thousand *Kan.*")

In the first movement, Chikamatsu sets the mood of the *michiyuki* with language such as "this fatal journey down the hill at Sakanoshita" and "hurrying on his pack horse through the late night." Movement B opens with: "We the sheep for slaughter, must wander through limbo seven times seven days with no stopping place," which explicitly suggests that the lovers are already dead. In the rest of this movement, Koman, and then Yosaku, recall their past lives. Koman describes her nine years as a courtesan, and Yosaku, the beginnings of their love. But their love will not be fulfilled in this world, and the final image is particularly striking, "We shall bury our fame in Sunken Fields." The direction of their journey moves ever downward. Movement C is sung in the *utazaimon* (ballad) style. In it Yosaku and Koman recall the former pilgrimage of love they made along the same road; the intensity of their love unites them. In movement D, the imagery turns morbid as they witness their own corpses, a "hideous spectacle," and their tombstones as well. The language also suggests that their former pilgrimage to the shrines portended their present journey through hell.

The language of this *michiyuki* is marvelously intricate. Chikamatsu's metaphor for the pitiful lovers, "a summer cicada that knows neither spring nor autumn," sets up a temporal sequence, spring, summer, autumn which is analogous to past, present, future. The lovers are looking backward from their present miseries, and this continues until the admonishing phrase, "No, we must not think of the past, never to return." In the Japanese original it is somewhat clearer, because three sentences in sequence all have *omoidasu* (recall) as the verb. The next line, about not weeping, is a play on words; it also means the "cawing of crows." At this startling sound of shrieking crows, they realize fully the proverb about being able to tell the moment of death from the cawing of crows. The language is striking, because the phrase "now I know that it is true" ("*ima zo shiru*") can also mean "to comprehend the present moment." After this the pair see their defiled corpses blocking the road, and at that moment the following line appears, "The past and future are both told by the present." This journey through the depths of hell and all its misery leads them to the state of mind described by the above line, that is, in Buddhist terms, a state of enlightenment. It is no accident that this moment is followed by the coming of dawn, the symbolic end of their dark journey. In the final movement, Koman envisions a happy future in the next life. The metaphor for this is "the Inn of a Hundred Delights, (where) Kannon and Seishi will take our hands to welcome us aboard a lotus leaf, an inn where you and I are the only guests." To complete this vision, death is necessary.

This "Road of Dreams" *michiyuki* is concerned only with the love-suicide theme, and as a consequence the story centers on Koman. The following section, which continues until the final dance, returns the pair from the dream world to the samurai world of Yosaku's past, the other theme of the drama. In this section, Yosaku again takes center stage and banishes all desires in preparation to die. His language is Buddhistic: "To banish these delusions we must free ourselves of the cycle of birth and death and pass into Nirvana."[22] Both Yosaku and Koman struggle to abandon their remaining cares in this world; for Yosaku this is concern for the fate of his son, and for Koman it is anxiety for her imprisoned father. In the end, they decide to die for these two: "for my father's sake, for your son's sake, let us plunge into hell!"[23] In the *michiyuki,* death is necessary for fulfillment of love; here it is atonement for those left behind. But, at this juncture, the pair are captured and their suicide aborted.

In the next scene, the *yatsushi*-samurai theme reaches its completion. Yosaku's encounter with his childhood friend Sanai is a revealing one. Sanai represents the ideals of the warrior cult, and foremost among them is the edict of absolute loyalty to the samurai's lord. He has come, as the messenger of the young princess (the daughter of Yosaku's former master), to tell Yosaku he has been pardoned and can be a samurai again. Though Yosaku is offered a new life by the young princess, he still wishes to die rather than live any longer with his shame. Sanai reprimands him from the standpoint of his former samurai status and admonishes him for his desire to die in a love suicide. For a samurai, the only honorable death is to die in battle or in an effort to save his lord. Sanai shames him by saying that, if he does not understand the way of the samurai, he is no more than a beast. Sanai's admonitions finally strike home and Yosaku gives up his desire for death with Koman. Sanai, then, gives Yosaku back his samurai status with the line, "You are Yosaku once more,"[24] and all the conflicts of the play are resolved—at least those that concern the theme of the fallen figure who eventually returns to his rightful position. However, the other theme, that of the love-suicide story, is left hanging, because the vision Koman experienced in the *michiyuki* cannot be fulfilled without death.

The content of the "Yosaku odori" dance, which brings the play to a lively end, has nothing to do with the "Yosaku" songs or dances current in Chikamatsu's lifetime. Why would Chikamatsu make the content of a supposedly festive dance the solemn tale of a love suicide? It is not unusual to end a *yatsushi* play with a dance of celebration because that is the convention; only the content is odd. We do not know how this final dance was staged, but, if we were to give any meaning to the dance in a production today, I would suggest that Yosaku and Koman take the roles of Tokubei and Ofusa, who are the doomed lovers in *Love Suicides at the Sunken Well,* the subject of the final dance. If we imagine the two performing this dance of death, then it is possible to suggest an interpretation of Chikamatsu's motives for including this subject as the finale. During the dance, the two lovers symbolically complete the love-suicide theme through the medium of art. Seen in this way the final *maki* appears more balanced; the dream dance of the *michiyuki* is completed by the final dance of death—a play within a play—and in between these two colorful scenes there is the solemn encounter with the austere world of samurai ethics.

An audience's expectations in watching a Kabuki (or Jōruri) *yatsushi* play are very different from the expectations in viewing a love-suicide play; at least one today would think that they should be. Yet, the play's popularity seems to suggest that the audiences of Chikamatsu's time saw no incongruities. Perhaps they enjoyed this drama, with its happy ending and mixture of styles, moods, and themes, as a *jidaimono* or a Kabuki-like play. They, of course, viewed it as theater, not literature.

THE UPROOTED PINE

Yamazaki Yojibei nebiki no kadomatsu (1718), translated by Keene as *The Uprooted Pine,* engenders less animosity among critics, because it does not have a mixture of contrasting genres evident in *Yosaku from Tamba.* It is Chikamatsu's finest example of a happy-ending *sewamono* play. In terms of the kinds of characters protrayed, it is similar to *Yodogoi shusse taki nobori* (A carp's successful climb up the waterfall), *Yūgiri Awa no Naruto* (Yūgiri and Izaemon), and Kaion's *Wankyū sue no Matsuyama* (Wankyū and Matsuyama); each has as its protagonists wealthy merchants: Yodoya Katsujirō, Fujiya Izaemon, and Wan'ya Kyūemon (or Wankyū), respectively. Each heir to an Osaka fortune fell in love with a courtesan of the highest rank and became so dissolute that he was disinherited. In the plays, at least, each of the figures is restored to his full inheritance in the end. The three high-level courtesans—Azuma, Yūgiri, and Matsuyama—were famous throughout Japan. The model is the Kabuki *yatsushi* drama. In particular the Yūgiri-Izaemon story, depicted in *Yūgiri Awa no Naruto,* was important to Chikamatsu during his Kabuki-playwriting period, because Sakata Tōjūrō, for whom Chikamatsu wrote most of his plays, gained his fame with the Izaemon role as his trademark. Many of Chikamatsu's Kabuki have a similar theme. The pattern has three acts, with the middle one taking place in the pleasure quarter, where the rich young man, now in the disguise (*yatsushi*) of a poor wretch, meets his lover after a long lapse. Each accuses the other of neglect or infidelity, both relate the troubles they have been through, and a tender reconciliation follows.

The woman figures in the four plays are all courtesans of the highest class in the pleasure quarter. Courtesans were of two major classes: *mise-jorō* (*hashi-jorō*), common prostitutes who populate the love-suicide plays and who are forced by poverty into the calling; and *age-jorō*, women trained for many years in conversation and in cultural arts such as music, poetry and dance. The *age-jorō* class, consisting of various gradations of rank, leads up to the highest *tayū*, the class of the women in these four plays. These women had the right to pick and choose their customers. The traditional geisha of modern times are similar to the *age-jorō* courtesans, because they are trained from an early age as entertainers, not as prostitutes, although they would be dependent on a patron or patrons for financial security. In Chikamatsu's time, few women reached the *tayū* level. Those who did were renowned throughout Japan for their charm and sophistication.

The Uprooted Pine begins during New Year festivities in the pleasure quarter of Osaka. Azuma, the reigning queen of the quarter, is a courtesan of the highest rank. An old woman approaches in an attempt to convince Azuma to meet briefly with her son Yohei, who adores Azuma, but who is too poor to afford a meeting. Azuma pities him and offers money, which he at first refuses, but finally accepts with the intention of using it to build a fortune and then repay her kindness by freeing her to become the wife of her lover Yojibei. Azuma also kindly gives him one of Yojibei's robes. The rich, drunken villain Hikosuke accosts Azuma, but is warded off by Yohei. Then Yojibei arrives and hears of Yohei's defense of Azuma; they become sworn brothers. In the final scene of *maki* i, Hikosuke mistakenly attacks Yohei, who is wearing Yojibei's kimono, thinking it to be Yojibei. But in the ensuing fight Hikosuke is stabbed by Yohei, who flees in order to make his fortune. Yojibei is captured instead and accepts the blame for the crime.

Maki ii is set at Yojibei's father, Jōkan's, house, where Yojibei is held prisoner. The father is a rich, miserly merchant; the father-in-law Jibuemon, a kind-hearted samurai. Jōkan is portrayed until the end of *maki* ii as extremely cold-hearted and unwilling to part with even a copper to save his son. Azuma comes surreptitiously to visit Yojibei, but meets instead an angry Okiku, Yojibei's wife. Okiku, however, finally won over by Azuma's resolve to kill herself for love, lets the courtesan into the house. In the final movement of *maki* ii, Jōkan reveals that he has in fact been trying to save Yojibei, and that he is willing to sacrifice himself by taking the blame for Yojibei's escape from the house. In the end, he forces Yojibei to flee, driving Yojibei—caught between Azuma and love of family—to madness.

Azuma and Yojibei's *michiyuki* opens *maki* iii. Theirs is a circular journey of madness with neither destination nor goal. The last part of *maki* iii returns to the quarter, where Yohei finally ransoms the missing Azuma, and brings the pair out of the trunk in which they are hiding. Yohei forgives and dismisses Hikosuke for his extortion attempts. The play ends on a festive note with Yohei spreading gold, sake, and good cheer.

The organization of *The Uprooted Pine* is opposite from a *yatsushi* Kabuki play. The settings of the three *maki* are: i, the quarter; ii, the father's residence; and iii, a *michiyuki* followed by a return to the quarter. The play's structure is symmetrical. *Maki* i begins in the quarter among extremely sophisticated courtesans. This enchanting atmosphere, however, is disturbed by the entrance of the comic-villain, Hikosuke, in the last half of the *maki*. *Maki* iii is the reverse of *maki* i: Hikosuke's banishment returns the play to the opening level. The middle scenes take place at Jōkan's house, Yojibei's prison. The form is symmetrical in that almost all the characters of *maki* i reassemble at the end of the play.

Chikamatsu certainly was aware of an earlier work by Kaion, *Wankyū sue no Matsuyama* (Wankyū and Matsuyama), which employs the same sequence of scenes: the quarter, a prison-home, a *michiyuki,* the quarter. Similarly, Wankyū goes mad during his *michiyuki.* In this play, the father banishes Wankyū in *maki* i to make him learn a lesson through hardship, thereby saving his life. At the end of *maki* i, Wankyū says that his first twenty-five years were a dream, and life in the quarter was like being in the Western Paradise.

Maki ii is a fall from that upper world of pleasure to the depths of suffering. The imagery depicts a vivid hell, complete with hungry demons. Wankyū's predicament parallels Yojibei's; both are caught between two women and the consequent responsibilities. In both cases, the wife in the end is sympathetic to the love between the courtesan and her husband. This sympathy and absolute devotion coming from two women—a wife and a courtesan—drives both Wankyū and Yojibei to madness. Wankyū is faced by a threat from both women to commit suicide if he does not go with the other. Wankyū is left alone; Matsuyama chooses to return to the quarter and allow herself to be ransomed by a wealthy country fellow, but after she has left, Wankyū goes mad for love and sets out to find Matsuyama.

During his journey of madness, he wanders about destitute. Matsuyama's new husband, Hayajirō, even offers money to the beggar Wankyū, but he politely refuses. Then Matsuyama, too, offers some money to him, and Wankyū dances a deranged, happy dance, proclaiming that in the end Matsuyama will be faithful to him. The suspense continues until Hayajirō explains that he knew all along the beggar was Wankyū. He gives Matsuyama to Wankyū, and even agrees to plead with Wankyū's father, Hayajiro's friend, to forgive the prodigal Wankyū. Love wins out over all; the play ends in unity and harmony.

In *The Uprooted Pine,* Chikamatsu places the beginning and end of the drama in the refined sphere of *sui*, an important term in the aesthetics of Tokugawa culture, meaning sensitivity and sophistication—*savoir faire* in matters of taste—far above that of the mundane world of money, violence, or evil. In this play, *sui* describes the world where the most elegant and refined courtesans rule by their delicate sense of taste, contrasting with the everyday life of the money-worshipping, merchant class. The encounter of Yohei and Azuma at the beginning of *maki* i is within this sphere; here we see the kindness of goddesslike

Azuma sow the seeds for the happy ending. The language is refined and devoid of any crudity, but, with the entrance of Hikosuke, the atmosphere suddenly changes. Azuma, whose "eyes flash love and wisdom" and who is "a courtesan in name only," is called instead "just a high-priced whore" by Hikosuke. His language is confused in other ways as well. He gives grievances instead of festive greetings at the New Year, and says such notable things as, "I proudly claim the honor to be an unrelated descendant of Emperor Kammu." He is, of course, a comic figure and his language reflects his role, but there is more to his language than just humor. It is violent, crude language full of earthy metaphors; the narrator says he stinks like "an overripe persimmon," and he refers to himself as a dog on all fours, after he is beaten by Yohei. Earlier the narrator comments on his action, "He drags her (Azuma) with the violence that drunkenness and his native brute strength supply." In contrast to Hikosuke, Yojibei is introduced as one who has surpassed the highest mountain of *sui* ("*sui no sui o koetaru koi no yama*"). Hikosuke is the agent who disturbs the world of *sui* surrounding Azuma, Yojibei, and Yohei. He stabs at Yohei, thinking that it is Yojibei, only to be wounded himself instead, and his accusations of murder against Yojibei throw Yojibei into prison. Not until the end of the play does the setting return to the sphere of *sui.*

The structure can be analyzed, as well, in terms of Gidayū's theory of a five-act play. The first part of *maki* i can be classified under the title of "Love," and the last section of *maki* i, with its fight and coarse langauge, fits Gidayū's Act II title of Shura (battles). *Maki* ii is the center of the play, and, in the pathos portrayed, contains the essence of the whole drama, as Gidayū says Act III should. The first scene of *maki* iii is a *michiyuki* journey, matching Act IV in Gidayū's theory. Finally, the auspicious conclusion completes the analogy with Gidayū's treatise on the form of a five-act history play. Although this is a tightly unified play thematically, each part has a different mood, and the variation is purposeful. Chikamatsu is using the conventions of the history play as the backbone of the dramatic structure of this contemporary-life play. Therefore, it is simplistic to say Chikamatsu was influenced by the Kabuki *yatsushi* genre, or by Kaion's earlier play; he uses elements from a variety of sources, and all are significant as background for understanding the style and conventions.

Chikamatsu is exact in his use of convention. The first two lines hint at the themes and outcome of the play. Each line is from a different poem. The first line comes from a poem in the *Kokinshū*, Book VII, number 350:

Kame no o no	Down from Turtletail Mountain
Yama no iwane o	Battering the rocks, tumble
Tomete otsuru	White jewels of the waterfall—
Taki no shiratama	How beautiful and endless
Chiyo no kazu ka mo	Is their infinite number!

The other poem is from the *Gosenshū*, Book XI, number 777:

Tsukubane no	From the peak of Mt. Tsukuba
Mine yori otsuru	As Minano River
Minano kawa	Tumbling down—
Koi zo tsumorite	My love, piling higher and higher,
Fuchi to narikeru	Forms a deep pool.

The first is a festive poem and sets the auspicious atmosphere of New Year. The second is festive as well, but is about love. Since "Minano River" can be written with the characters for man and woman, the image of the water rushing into a deep pool is quite suggestive. In both poems, abundance is important, whether depth of love or number of jewel drops. Multiplicity continues in the lines that follow the opening of the play: the shuttlecock goes back and forth many times, and the array of beautiful robes are all "double" or "triple-dyes." This fascination with numbers appears at significant moments throughout. As Yohei takes the few coins that he received from Azuma, he says, "I'll turn these ten gold coins into a hundred, a hundred into two hundred, two hundred into five hundred—once the profits start rolling in, I'll turn my fortune into a thousand pieces of gold quick as shooting an arrow."[25] This atmosphere of abundance and multiplicity, however, is shattered when the first crisis occurs, the wounding of Hikosuke. But at the very end of the play, when the money for Azuma's contract is paid by Yohei, who has returned from Edo after making his fortune, the numerous joyous handclaps

CHART 11
Musical Structure of *The Uprooted Pine, Maki* ii[a]

A. Okiku's anger and sympathy for Yojibei, now confined
   1. Description of Yojibei's situation, "quarters unvisited even by the sun." (to Keene, p. 328 "Yojibei helplessly awaits what the morrow will bring.")
   2. a. Okiku's concern for welfare of distraught Yojibei (*noru fushi* p. 328 "stares blankly at the floor in utter dejection.")
      b. (*kotoba*) Okiku's bitterness and jealousy (*suete* p. 328 "Bitterness is mixed in her tears of indignation.")
      c. (*kotoba*) Yojibei's determination to protect Yohei (*fushi* p. 329 "He bows his head, a piteous sight.")
      d. (*kotoba*) Okiku's anger at Jōkan who is so miserly (p. 329 "made his heart all the darker.")

B. The chess game
   3. Jōkan is trapped by Jibuemon in chess (p. 330 "He ponders the situation.")
   4. Jōkan remains steadfast in refusing to give money for Yojibei's life (p. 331 "it seems as though Yojibei's life has been checkmated.")
   5. a. Jibuemon releases his anger at Jōkan for his stinginess (*noru chū* p. 332 "Jōkan blinks his eyes for the tears.")
      b. (*ji haru*) Jōkan's feelings, "The more affectionately I think of him, the harder I find it to give him the money." (*suete* p. 333 "He leans his shaven head over the chess board and weeps.")
      c. (*ji chū*) Jōkan's feelings turn to frustrated anger. (p. 333 "unable to speak a word of reproach, goes outside weeping.")
   6. Okiku's anger at Jōkan's miserliness (*sanjū* p. 333 "and the darkness spreads.")

C. Azuma's arrival and encounter with Okiku
   7. a. Azuma's journey to Yamazaki (*fushi* p. 334 "that this is his prison.")
      b. (*kotoba*) Azuma alights (*fushi* p. 334 "the closer she gets, the higher it seems.")
      c. (*chū u*) Azuma cannot sneak in (p. 334 "her body chilled by the bitter cold.")
   8. a. Azuma mistakes Okiku for Yojibei and gives love letter with razor (*fushi* p. 335 "She opens the garden door and comes out.")
      b. (no notation) Encounter of Azuma and angry Okiku (p. 336 "His only wish is to vanish altogether.")
   9. Azuma shows her determination to die (p. 337 "yields to tears of regret at leaving this world of dust.")
   10. Okiku relents (p. 337 "their reserve melted.")

D. Jōkan forces Yojibei to accept mercy
   11. Jōkan's analogy of the mousetrap (p. 338 "Tonight I will be able to read the sutras with a mind at peace.")
   12. a. Okiku begs Yojibei to obey and leave (*suete* p. 339 "Okiku catches the note of urgency in her words.")
      b. (*ji iro u*) Yojibei refuses to leave and endanger father (p. 339 "they can only weep.")

13. a. Jōkan bares his true feelings (*fushi* p. 340 "the more painful it has been.")
    b. (*ji u*) Jōkan threatens suicide (*suete* p. 340 "He collapses in tears.")
    c. (*kotoba*) Jōkan forgives Yojibei's past wrongs (p. 340 "the rest is choked by tears.")

E. Yojibei falls into madness and leaves Azuma
  14. Yojibei loses his senses, and they depart.

Note: [a]Capital letters mark movements; arabic numerals, primary units; and small letters mark subdivisions within primary units. Page references are to Keene, *Major Plays of Chikamatsu.*

of the characters, flowing gold coins, many festive toasts, and nuptial pledges "that through a thousand years, ten thousand years of life,"[26] promise fidelity—all return the atmosphere to the auspicious multiplicity and harmony that began the play. Hikosuke is barred from this final scene by Yohei, who pardons the villain, but banishes him saying, "since this is a festive occasion—leave at once!" However, if the play were only to offer a happy vision with no crisis, there would be no dramatic story to enact. We must look at *maki* ii and the *michiyuki* to see how the playwright dramatizes the crisis of Yojibei's fall from the world of *sui* to his prison-hell and madness.

Yojibei falls into a world "unvisited even by the sun" and into a voluntary dreamlike exile. Yet this world is none other than his own home, where he is among his family. In this prison, his suffering is caused by the absolute devotion of those around him rather than by any evil. The figure who dominates the action of this *maki*, however, is not Yojibei, but his father Jōkan, the miserly millionaire. He is very near in personality to the character of Wankyū's father in Kaion's play, because each attempts to save his only child, and the revelation of their feelings is the climax of the drama.

The *maki* is balanced between dramatic scenes involving Okiku and Azuma on the one hand, and Jōkan on the other. Yojibei has only a supporting role. There is little action; dramatic conflict is confined within the noble heart of the elderly merchant Jōkan. Although he is trying to help his son by remaining steadfast in his refusal to pay a ransom to Hikosuke, he is forced to bear the indignation of Okiku and her father Jibuemon. His true feelings are gradually revealed, and the climax of the revelation is in movement D. Since his honor demands that he cannot openly help his son escape, he uses a variety of indirect means to achieve his goal. During the chess game, his recital of the passage from the Nō play *Kanehira* describes both a warrior sinking into the Buddhist Shura realm, and, indirectly, Yojibei's fall into trouble. The mousetrap story is another of Jōkan's roundabout attempts to persuade Yojibei to escape. Jōkan uses the verb *ochiru,* "to fall down" ten or more times: in the Nō passage, the horse falls into the mire; in the chess game, when the opponent loses he too falls; and, finally, in the mousetrap analogy, the verb "to escape" is the same as "to fall." Jōkan steadily pushes Yojibei until he agrees to "*ochimashō,*" to escape from the house and fall into the world of freedom, which for him is madness.

The action is somewhat perplexing if viewed as strictly realistic drama. All the characters are extreme in their undying, absolute devotion to the frivolous Yojibei, who, though not a bad man, is hardly a commanding presence. With each exhibition of love from those around him, Yojibei is driven further towards insanity. The crowning point is the moment in unit 13, when Jōkan forces Yojibei to flee on the threat of Jōkan's own suicide. The action of Okiku lifting "the earth trodden by her father-in-law's feet" suggests a view of Jōkan as an Amida-like figure, who offers absolute mercy and freedom for the sinner. Let us look more closely at the passage in which Jōkan demands Yojibei accept his mercy and consequently pushes him into a hell of madness. (See Glossaries B and C).

ji haru
13a Jōkan calls from inside.
iro *kotoba*
(Jōkan) Okiku, Okiku! Do I hear aright? Is that disobedient scoundrel saying that he won't escape? Heartless, inhuman wretch!—I haven't told anyone for fear that people might say that old Jōkan at the age of seventy allowed himself to be blackmailed, but I secretly sent a man to Hikosuke. I tried to settle things privately by offering up two hundred pieces of gold. Hikosuke took advantage

of my position, and said he was unwilling, even for a thousand pieces. I've heard that his wounds are superficial, but he's mortal, after all, and there's no telling when he might die. Can you imagine the agonies I've gone through as a father? *ji iro chū* I have tried to distract myself with chess, *haru* only to be harassed. *chū* The more I have been reminded of Yojibei's plight, *fushi* the more painful it has been.

b *ji u* All evening long I have racked my mind *haru* thinking up that plan of *ue* escape from the trap. Try to understand the *u* immeasurable fatherly love that went into it. (To Yojibei) You cannot remain a child all your life. Some day you'll be a father yourself. If only I could make you realize how a father feels for his son! Tell me at once—will you or won't you escape?

(Narrator) *kin* His voice is harsh, but the *haru* tears on his face may be surmised even by those outside the wall.

ji chū Yojibei, weeping, *iro* prostrates himself.

(Yojibei) *kotoba* My gratitude makes it all the more impossible to run away. *ji iro u* Forgive me, I implore you.

(Narrator) He bows his head in tears.

(Jōkan) *kotoba* Very well. A man normally hopes to survive an aged father, even by a single day, and he tries to remain in good health so that on the anniversaries of his father's death he may offer prayers for his happiness in the other world. *ji iro haru* Would you prefer *u* to be mourned by your father and make him suffer? Very well, stay here if you

like, but first I will thrust this dagger into my wrinkled belly.
u
Namu Amida Butsu.
ue *u* *suete* *u*
(Yojibei) Father! I'll go away! Wait! He collapses in tears.

kotoba
c (Jōkan) Do you swear it?

(Yojibei) How could I lie to you?

(Jōkan) I'm happy. My mind is at rest. I forgive you all your dis-

obedience of the past. I accept this one act as the equivalent of
ji haru
thirty years of dutiful service as a son. Your dead mother must also
u *iro*
be rejoicing. Okiku has a father, and I have Okiku. Don't worry
kotoba
about us here.—I see that you'll have a lady to go with you. Madam,

please make sure that Yojibei takes moxa treatment however much

he objects, and don't let him drink any sake. Always travel in a

sedan chair, even if it's expensive—people may recognize you on
ji iro chū
horseback. Please do as I ask.

(Narrator) His words are brief, but his agitated heart is torn in ways
u
numberless as the stripes of the purse he tosses to Yojibei. "Fare-
fushi *noru*
well" is his only word; the rest is choked in tears.[27]

The musical notation outlines the rhetorical methods used in performance. Jōkan describes his feelings and confesses his attempts to bribe Hikosuke. His monologue breaks at the emotional point when he describes the harassment by the others around him, while he was trying to be calm and think. After the relaxing *fushi* cadence, his monologue continues. In section b, he forces Yojibei to accept his mercy even though it endangers

Jōkan's own life. The beginning, sung in the adorned *ji* style, employs a very high pitch (*haru, ue*). The narrator's comments after his imploring speech are in the emotional, high-pitched style of *kin,* followed by a gradual lowering of the pitch (*haru, u,* and *chū*). Yojibei's voice, spoken in *kotoba,* relaxes the emotional build-up. When Jōkan next speaks, he declares his determination to die before his son. His voice, too, is in rational *kotoba.* The second half of his pronouncement shifts to *ji iro* as he makes a final impassioned plea before killing himself. This emotional entreaty is interrupted by Yojibei's outcry in a high-pitched voice (*ue*). The harsh *suete* partial cadence brings section b to a close. Jōkan's next lines are in *kotoba* and the mood is suddenly light and the pace quickened, as he tries to hurry Yojibei out of the house. The final part returns to the emotional song style for the sorrowful parting.

Jōkan's act of mercy pushes Yojibei into madness. Yojibei's agony comes, ironically, from the love and mercy that surround him. Each character is resolute in his behavior and attitude towards Yojibei; he alone has no excuse. Although his prison is the comfortable womb of his own home, this total love and freedom force him into a hell of alienation. The cause of Yojibei's dilemma is overtly Hikosuke's villainy, but the real problem within Yojibei's heart is unrelated to that incident. The conflict between his responsibilities and his desires for juvenile freedom already exists. Hikosuke's action is merely the catalyst that brings it to a critical impasse; Yojibei can no longer waver indecisively between these two poles. He is forced to choose either family or desire, and ultimately he must choose desire.

The *michiyuki* of Azuma and Yojibei differs from a love-suicide *michiyuki.* Theirs is a mad journey without destination—only away from Yamazaki. Although it is a journey of suffering, the imagery does not imply a journey through hell as the love-suicide *michiyuki* does. The scene divides into five movements.

A. Introduction: the sadness of two who cannot know the other ("What a sad contrast to Yojibei!" p. 342)
B. Song of madness: ("She places a staying hand on him." p. 342)
C. Yojibei remembers and regrets: "I am an autumn leaf

careening madly of itself, too unsettled to alight." ("Her last words to me will be my strength." p. 343)

D. Despair and longing for father and wife: "My heart is one but torn in two," ("They take each other's hands and weep, giving voice to their grief." p. 344)

E. The circle of suffering: *michiyuki* ends with dusk.

Structurally this *michiyuki* is similar to those in love-suicide plays in its movement to a crisis, followed by a denouement. There is, however, no suggestion of hope at the end; instead, there is a vision of endless suffering, and moreover, having the journey end at dusk reinforces the atmosphere of despair. Movement D is the climax. In it, Yojibei is driven by thoughts of Azuma on the one hand, and his wife and father on the other, to the point of having his heart "torn in two." This split vision of suffering and alienation is at the opposite pole from the harmonious and unified vision of abundance at the beginning and end of the drama; and it is the experience of the fall into alienation and madness that leads to the happy ending.

Hikosuke's two comments in the last scene of *maki* iii, that he heard Yojibei had dropped dead along the road, and that he will kill Azuma with mercy, are, on the surface, only Hikosuke's typically derisive comments. They do, however, suggest the other theme, Yojibei's rebirth as a new man after his period of torment. It is in our expectations of a Yojibei reborn with a new awareness of the world that we are disappointed. Yojibei does appear at the end of the play and does participate in the festivities, but he and Azuma say hardly a word, offering only humble apologies. We are given no hints of what the "new" Yojibei is like. Yojibei's experience, which had been very personal in *maki* ii and during the *michiyuki* journey, suddenly becomes public, and we are not given access to his inner feelings. What we do see in this play (and in other late works), however, is Chikamatsu's masterful presentation of the psychology of elderly parents. Jōkan's inner conflict—rather than the dilemma of the young people—is at the core of this drama.

In a history play, the subject is the public world, but in the contemporary-life play usually the private sphere of common men and women is in the limelight. This *sewamono,* however, ends as a comedy in the public world with public forgiveness for

Azuma, Yojibei, and even Hikosuke. The play comes full circle and the audience is returned to the *sui* world of harmony, each with his own private vision to take home from the theater.

Chikamatsu's many important dramas offer audiences, today as in the past, either as spectators or readers, interesting, exciting, and rewarding experiences. For most, however, Chikamatsu's stage is far away both in time and space; and his plays different from those familiar in the West. Our understanding and appreciation of his works can be enhanced by viewing his dramas from within their tradition. From without, the view will always be clouded. His plays cannot travel to our world; we must journey to them. An awareness of the conventions of performance brings us closer to his stage, and allows us a clearer perspective—from within his circles of fantasy—where we can reach out and touch the multisensual textures of his theater, and savor its delights.

Appendixes

Appendix A

Preface to *A Collection of Bamboo Shoots* (1678)[1]
by Uji Kaganojō

In Jōruri, there are no teachers.[2] However, one should understand that its parent is the Nō.[3] The reasons for this are first, at the beginning of a day's performance of Jōruri, the puppets perform the Nō dances Okina, Senzai, and Sambasō. Then, when this ritual has ended, an opening melody from Nō is played as the puppets make their entrances. Are not the robes and costumes for the important roles the same as in the Nō? Therefore, the source of Jōruri is the Nō. The dramatic methods and techniques of Jōruri are neither too subtle nor too overbearing. One must chant each piece in a style suitable to its mood. Since all the many paths of art are born and nurtured from the heart, one should follow the dictates of his heart. However, since the novice will most likely be unable to comprehend this without assistance, I will give an outline of the methods of the art of Jōruri.

The basic scheme lies in these four musical styles: the auspicious (*shūgen*), the elegant (*yūgen*), the amorous (*renbo*), and the tragic (*aishō*).[4] In an auspicious piece, the sound of the words, when chanting, must be elegant and proper; that is, not too heavy nor too light, yet they must be delivered with vigor and strength. This must be the foundation for all musical styles.

In chanting other styles, the voice should always be the same as for the auspicious one, while deep in the heart the feeling behind the voice is altered, according to the nature of the piece.

For the elegant style, the voice should be the same as when chanting auspiciously, but the song should be softer, and the heart should feel as when one is sitting under the blossoms, listening to the koto, flute, or *tsuzumi* drum while passing the time of day, forgetting about the journey home.

As for the amorous style, the effort of the heart must be more fervent and earnest than for an elegant piece. One must be moved by the lovers, have a yearning heart, and chant with deep concentration. When one sinks deeply into the mood, the color of that mood manifests itself naturally. The clear dew, when it settles on a red leaf, becomes a crimson jewel.

Finally, there is the tragic style. For this, all the previous sentiments must be forgotten, only a lingering touch of love should remain. Yet the heart must maintain its strength and in its depths should lie the feeling of life's uncertainty, of its transiency.

In performing any piece, posture is always most important. First of all, in the kneeling position legs should be apart, the hips set firmly back on the legs, and the head held correctly upright. One's heart should remain relaxed, and concentration should be centered on the spine. When the posture is poor, the voice and performance is poor. This is true for any art. If anyone does not understand this, it is because his accomplishments are meager.

As for the music of Jōruri, there are four essential elements. First is the pitch; second, the rhythm; third, the melody; and, fourth, the timing [meaning adjustment of the mood to the occasion or to the feeling of the piece]. One must always consider the particular musical style when chanting. These are the four essential elements of the music.

The melody maintains the style; the style maintains the melody. Also the beat maintains the pause; the pause maintains the beat. This must be well understood.

The art of chanting the musical narrative of Jōruri makes use of the mouth, tongue, and heart. The mouth and tongue listen carefully, and express what the heart feels. Among these elements, no one should take preference over the others. For example, even if the voice is suitable, if one uses words inappropriate to the particular audience, for instance, at the house of a lord, or forgets one's own special melodies, the performance will not be entertaining. This must be well understood.

The mouth must study and master the syllabary so that it can smoothly and distinctly recite the text; each syllable must be carefully enunciated, yet not be stressed too much. To chant mellifluously is the function of the tongue. When the heart knows the laws of the universe, this understanding forms the basis for the four musical styles and the four elements of the music. It is only the heart that can bring art to fruition.

In all music, one must refrain from exerting force into the voice. One's strength must, instead, be held in the heart. This certainly is true for Jōruri.

> Music is like a grand bamboo
> Straight and pure with few nodes.[5]

This poem shows a clear understanding of music. Although there should be few intonations in the melody (*fushi*), the singing must not be feeble. A certain amount of distinct intonations are necessary. On a healthy bamboo too the nodes (*fushi*) are pleasant to behold, while on sickly bamboo the haphazard nodes are unsightly. Likewise, if you take the branches of a garden tree and fashion them into various shapes, it makes them interesting, but soon they lose their charm. However, deep in the mountains near the rocks where the small pines naturally receive the blessing of rain and dew and gradually become covered with moss—such trees are so indescribably magnificent that their beauty and charm penetrate to the depths of one's soul.[6] Jōruri too will retain its flavor, if the four musical styles and the enunciation of the syllabary are carefully studied, and if the text is recited to traditional melodies. However, being unable to keep to classical standard, I mix other music, such as popular songs and ballads, into Jōruri, which makes the musical product the same as a sickly bamboo or an artifically-fashioned tree. Those who have discerning tastes will naturally find this music strange. Since such people are scarce, however, I must follow the whims and fashions of the times, although I myself consider them base.

If one does insert other music into Jōruri, since the transfer of other melodies to Jōruri is like grafting bamboo onto tree branches, one must practice again and again to make it harmonize with the whole.

Each of the four basic musical styles of a play must be chanted differently. When one does not understand these various musical styles and cannot distinguish them but concentrates on the melody alone, when a chanter lets his rhythm be taken over by the samisen, or when he becomes slack and careless, his concentration has slipped away. Therefore, if one allows one's posture to collapse, the voice too will gradually weaken. Those who do not fully understand these matters will not be in control when chanting, but will rather be controlled by the text. Jōruri is the needle, and the samisen, the thread. This must be well understood.

In the Nō, there is a special training for enunciation of the syllables *no mo to shi mo yo no ya mi mu.* Although this should be part of the training for Jōruri too, it is rare to find one who has mastered the art of chanting to such a degree. Most chanters strive only to keep too harsh or too delicate sounds from striking the ear.

These days, if one watches the chanting of Jōruri on stage, he will notice that chanters often turn away from the direction of the audience or even turn their backs to it.[7] This sloppy method of performance is naturally impossible for me to condone. When there is a disturbance that distracts one's attention, should one let this become a hindrance to art? Even in cases where one is disturbed during a performance, if one retains concentration and devotion then the performance will not be hindered. As I explained above, the posture must be disciplined. When one does not release the voice from the stomach or trunk of the body, the face appears strained and its shape contorted. Those who follow my school must first of all clearly understand these principles. No other art is considered as base as Jōruri, because the performance of most chanters is crude. But already a few able Jōruri performers have been appreciated and have received honorary titles and offices from the court.[8] In any type of music, is not this the case? However, those who do not understand this, or those who are not even aware of the fact that there are four basic musical styles and four essential elements, and continue to chant carelessly, are poor artists indeed, and their work is crude.

On the meaning of *ji, ji iro, kotoba, fushi* and the musical notation of Jōruri[9]

In *ji* [song] passages the chanting follows the beat of the music. In the *ji iro* [between speech and song, parlando] passages the chanting does not follow the beat of the music. Therefore, because one does not follow the samisen melody, one must be very careful. The distinctions among the four musical styles must remain clear, and one must not let the recitation become sloppy. The long and short sounds should be released as if from a taut bow, but they must neither strike against the samisen, nor passively follow its rhythm.[10] Transitional passages must be chanted carefully and effectively, and each of the four musical styles must be considered while reciting. The above outlines *ji iro*-style chanting.

Kotoba should be recited as in the usual manner of telling a story and the four musical styles should be followed, but there should be no melody at all.

As a general rule, each chanter should develop melodies (*fushi*) which suit his own heart. This method alone should suffice, but since many have expressed a desire to learn more about it, I will explain briefly what I have learned during my years as a performer.

		This notation is like that used in Nō.
		This means to continue the melody.
		This means to continue with an enlivened spirit.
		This means to raise the voice.
ウ	*u*	This means to let one's feelings float (*uku*) with the music.[11]
色	*iro*	This means that the passage is neither natural speech nor song.
スエテ	*suete*	This means that the passage, while being melodic, is transitional between two styles of chanting.[12]
ヲクリ	*okuri*	Perhaps the feeling is similar to the transitional melodies of a Nō play. This also means that it is a transitional passage to the next musical section.

三重	*sanjū*	It need not be explained. However, one should consider carefully the demands of the particular musical style as well as the synchronization of one's heart with the movement of the dolls.

Besides these, there are other kinds of notation, but they need not be explained. The melody is really quite easy to chant, but strict training and discipline are necessary to master the four musical styles, the long and short sounds, and the transitional passages. Correct pronunciation and clear enunciation also require years of training.

Since the beating of a fan to keep the rhythm is used only to inform the samisen player of a transition to a new passage, it need not be struck at each pause or breath. Its use should be limited to hastening or slowing the rhythm of the transitional and following passages.

The above suggestions are generally applicable for those who chant at private residences. When puppets are used, minor changes are necessary. As with all the arts, nothing is as difficult as following the dictates of one's heart "out of the darkness, even further into the path of darkness,"[13] as the saying goes. This phrase seems most appropriate. Although I myself hardly know anything about the path of art, I have relied on the writings of the sage of Nō, and, while hoping that my clumsy efforts might be worthwhile, I have exhausted my mind to produce these writings and now offer them to you, my disciples. My audacity in writing this preface is truly like that of a raven who imitates the cormorant bird, tasting not the water of the sea but swallowing instead the slanders and criticisms of society. I regret divulging these secrets of the tradition.

Appendix B

Preface to *The 1687 Gidayū Collection of Jōruri Scenes*[1] by Takemoto Gidayū

The courtiers of the Palace decorate their hair and hats with blossoms in their leisure as they enjoy spring's glorious beauty.[2] Each person has his own preferred aesthetic pleasures, and Jōruri too can be ranked alongside the flowering arts of the nobility. However, even among flowers and plants, there are old trees as well as young ones. The blossoms of young cherry trees have bright colors and delightful fragrance in abundance, around which butterflies, birds, and passers-by frolic in joyous play. At the break of dawn after such a night of play, the bedroom suddenly becomes softly white, causing one to believe in surprise that the morning sky has come. At such moments, the light from the moon reflected in the young blossoms penetrates most intimately into the heart, giving one a feeling of joy in nature's radiant beauty. As time passes, however, even a famous and vigorous tree grows old; its color fades, its fragrance grows faint, and its heart yearns only for a return to the glories of its past. It is sad to see such a tree lose its charms.

In this land of many islands, for the art of *waka* poetry, too, is it not said that when one has both the flower [words] and its fruit [heart] together in equal parts, the product will be an excellent poem?[3] This principle is also true in Jōruri. However, to create art in this day and age, when one has the seed with which to entertain and make one's audience happy, it is most effective to chant making the flower [words] seven-tenths and the fruit [heart] three-tenths of one's effort.[4] Even more than the double-blossomed cherry trees or the thickly blossomed Tigertail cherry trees, the bright and delicate blossoms of the single-petal Yoshino cherry trees attract the most passers-by, who gather under their colorful shade. To be sure, the finest trees are the old ones deep in the valleys, upon which only a few scattered flowers quietly bloom, but, if one cultivates a young tree in imitation of these old ones, then its branches will not grow strong; rather, they will be stifled, and the blossoms will be feeble with little charm. The style of Jōruri too should be lively and vigorous, like the colorful image of the dandy, the fashionable and confident man-about-town. As the years have passed by, I have tried to maintain this lively spirit as the foundation of my art, but, although I still feel vigorous as in my youth, my style must now be old-fashioned and feeble, and my performance surely is no longer entertaining for today's audiences.

Recently, on a certain evening, an old and intimate friend came to visit, and asked me to instruct him in the secret teachings of the art of Jōruri. I answered in the following manner. Basically, since the art is modern, there are no traditional teachings to be learned from anyone. The music is restricted to a five-tone and twelve-note scale. The traditional songs and melodies of Kagura sacred music, popular songs, folk songs, Kōwaka-mai ballad drama, Nō, and Heike recitation are all incorporated into Jōruri; this is the tradition. If an artist performs well, is praised by others, and is then acknowledged by the world as a master, his words will certainly become secret teachings. However, if he slanders others while saying that only he himself is good in the hope that he will be considered a great artist, then for such a person there is the refined treatise, *Kadensho,* by the sage of Nō, which contains the secret precepts of the tradition, handed down from generation to generation.[5] In this work, the secrets of music are concisely explained. It contains many teachings about maintaining balance between the masculine [skill] and the feminine [heart], about rhythm, about enunciation and lengthening of syllables, about the four styles and four elements of music, and about musical notation, and so forth. These aspects of music are all explained in detail.

It is only natural that those who teach others about the art, or are acknowledged by the world as accomplished artists, must be familiar with these matters. However, to use quotations from the *Kadensho* in order to brandish pompously one's authority is distasteful. Those who skip their practice sessions and try to learn solely from secret teachings will never learn to perform well. Reading books on theory will neither help one's mouth to move smoothly nor one's voice to improve and become entertaining for the audience. As he practices constantly and becomes skilled, his own style and method will conform to the tradition. As has been stated in the past, true talent must resemble former and contemporary greatness, and yet be unique.[6] Although at the lofty Emperor's Court there are disputes about theories of art, even there the closely guarded secret teachings of poetry are of little use when composing poems; they can only reassure one's heart, as Konoe Motohiro of the Court has said.[7] To gain proficiency in Jōruri or any other art, one must practice and study day and night to master the essential principles. Only after many months and years of diligent training will the art begin to become one's own.

It has been said that Jōruri should consider the Nō its parent, and so one should learn Nō chanting first, and only then begin to practice Jōruri. My reply is that, although such an idea has its apparent merits, the artists who follow my style regard the former masters of Jōruri as father and mother, while considering Nō, Kōwaka-mai, and other traditional music as foster-parents. Children, however, do not know the hearts of their parents. Each individual has his own talent, his own feeling for the rhythm between the voice and heart, his own sense of poetics and of the rhythm of the samisen, and his own sensitivity to the reception by and criticism of the audience, which can never be completely learned from a teacher. In any case, one should work hard to make the chanting enjoyable and pleasing so that it does not strike the ears harshly or startle the listener. One must not stray from the essence of Jōruri, while striving to coordinate the narration so that it synchronizes smoothly with the movements of the puppets. When asked to perform at a private residence, one must tailor the performance to the desires of the patron. The ability to entertain without boring one's audience should be considered the secret tradition of the art of Jōruri. Those who achieve this skill should be considered masters.

Although Jōruri is not as great an art as Nō, what is the rhythm of the steps of the god in the Nō play *Takasago* for? Is not its purpose to entertain the hearts of the audience! Our teaching is to listen to many kinds of music, drama, and storytelling, and to discard that which is not pleasing to one's heart; that which remains will most likely be effective as art. One should adhere to the golden writings of the sages, while gathering familiar childhood songs and any melodies that one finds particularly pleasing, and use them to refresh and expand one's art. One must open one's ear and mind, because no one school has the secret teachings and traditions, handed down from generation to generation. One must think that one can oneself become a teacher of Jōruri. One should not, however, seek to become famous; instead, practice with full concentration to become proficient and skillful. If I consider the above to be the essentials of Jōruri, and if I then call this my secret teachings, there will be both those who are impressed and those who will laugh. When one enters into this path and tastes its charms, one will never know the limits of its exotic flavors, and even the sweetest honeydew will taste mild in comparison. I have exhausted my mind and brush! Ah, for the company of a friend!

AN OUTLINE OF THE STRUCTURE OF A JŌRURI PLAY

Act I: Love

From the opening lines to the end of the prelude is the Sambasō section of the program. It should begin in the *utai, ji,* or *fushi* style of chanting.[9] The choice of the opening prelude depends on the feeling in the piece. The pitch is usually on the first note of the scale. One should chant vigorously as if handling loose and wild reins, and strive to excite the feelings of the audience. The final scenes of all five acts are important, but the final scene of the first act is particularly crucial.[10] The response to the first act on the opening day is of utmost importance to the success or failure of the play. Love scenes and secret rendezvous have traditionally been included in the first act. Since the creation of pleasant appearances should be the primary objective of this act, while one's heart must be strong, the words should be light, and one's spirit gentle. One must chant the lines of the female characters so that the audience finds them pleasing to the ear. When one chants the love scenes tenderly, the audience will be charmed. The *serifu,* speech, parts are like the shallow rapids of a river; the *fushi,* melodic, passages are like quiet pools.

Act II: Shura *(Battles)*[11]

In the second act, the mood of the first act must be discarded, and one must chant without any touch of heaviness. The mood is light. On the style of the battle scenes, there is the traditional teaching of the late Harimatayū, who was famous for his vigorous rhythms and fast beats. One should chant with speed and force, so that the audience listens tensely without loosening their tightened fists. The pace of the rhythm should continuously vary from slow to quick and from quick to slow. In dance or violently argumentative passages, if the voice moves quickly, then the heart must remain calm. In the passages where the voice chants heavily, one's heart must float lightly.

Act III: Pathos

The third act is the heart of a Jōruri performance. If it were Nō, then this act would be the *kuse* section.[12] Within the structure of Jōruri, however, the content and style of this act are not exactly prescribed. Nevertheless, one must determine the mood and chant the third act accordingly. The mood of the third act is special. I will discuss later the ruler-subject relationship of the chanter to the samisen player. When chanting scenes of grief and sorrow, one must not disregard reality. While chanting this act, the essence of the entire play must be held within one's chest. The extreme voices of grief must be clearly expressed. The various voices of pleading and storytelling, of past and present, and of tears and sorrow must be clearly delineated in the delivery of the chanter. There is the oral tradition for this.

Act IV: Michiyuki *Travel Song*

One should extend the pauses and chant more gently than in the previous act. In this act, the entire play begins to come together as a whole, but since the feelings of the audience are exhausted after the third act, one must be careful not to tire them any further. The melody is the most important aspect of the *michiyuki* scene. There are all kinds of *michiyuki:* those of courtiers and peasants, old and young, men and women, long journeys and day trips, boat excursions and those along mountain paths, but for the chanter the only difference is in feeling. One accompanies the samisen and chants softly. As for the three musical tempos of *jo* [introduction], *ha* [intensification], and *kyū* [quick conclusion], this act is either the *jo* of the *kyū* or the *ha* of the *ha;*[13] there are conflicting oral traditions. In any case, one should at times accompany the samisen and at other times chant alone; at times follow the rhythm and at times chant against it. One must use all one's resources and skills.

Act V: Conclusion

Since this act must conclude the whole play, it is a difficult section. However, the fifth act in particular follows closely the Jōruri structure.[14] The mood is auspicious. Metaphorically, the first act is the silk fabric; the second, the silk lining; the third, the pattern, color, and embroidery; the fourth, the cotton padding; the final act, then, is the tailoring of the kimono. Since all of the words of the dialogue are part of a drama, it is vital that the language of the aristocrats, samurai, farmers, townsmen, villains, women, and the oracles of the Buddha and gods must be distinct and suitable to the individual character speaking.

Although one can create theories on the principles as well as on the techniques of art, neither Jōruri nor any other art is limited to them. In essence, Jōruri is a one-man mimic show. The chanter must be able to produce with his voice the sounds and sights of birds and beasts, of rippling water and rustling wind, and of things both with and without souls. However, I do not at all advocate recklessly or forcibly imitating the sound of a cow or a trumpet shell. There are two aspects of imitation. One is imitation of the action; the other is imitation of the spirit.

One's own judgment must be used to find the most effective balance between these two kinds of imitation. This is my teaching.

There is rhythm that follows the samisen and there is rhythm that goes against it.[15]

one way is to follow the samisen rhythm at this point

"Deep in thought, now again . . ."

another is to chant against it

Learn this well.

There is rhythm that is vertical [narrow] and horizontal [full or wide]. It is like a woven garment; therefore, the voice is the design.[16]

tate nobetaru kokoro
"Lately grown accustomed to the salt winds,
shin no kiri
While deep in thought, now again
nuki chijimu
They strike harshly in my ears, piercing into my body;
tate
It will be here [my love meets his death]."[17]

There is the lord-subject relationship; Jōruri is the lord and the samisen, the subject. In chanting, too, there is a style of lord-subject.

"Early maturing rice, late rice
shin no fushi
The average are all cut, as cards in a game
kun no fushi
one, two, three"

kun no fushi
"If one follows the trail of smoke,
shin no fushi
The eye alights on the sea of Ise, widely spreading."[18]

Su-Jōruri [unaccompanied Jōruri]. It is a mistake to say that this means chanting with the samisen but without the puppets. Su-Jōruri really means chanting without the samisen accompaniment.

The secret tradition of four *bu* [inches]-six *bu.*[19] One method is for a piece that is one *shaku* [foot]: of the first five inches, chant seven, and of the last five inches chant three; or of the first five inches chant only 2.5 inches and of the last five chant 7.5. As for the accompanying samisen or other musical instrument, the pauses of the samisen are maintained by the Jōruri chanter. Perhaps it is the same as the chanting of Nō with only one small *ko-tsuzumi* drum.[20] This is the secret teaching within the secret tradition.

Ji, kotoba, fushi. Ji is like the water; *kotoba* is the current; *fushi* is the quiet pool. This is part of the oral tradition.

> The aspect of visual expression, expression of feeling.
> The six techniques.
> The aspect of oral delivery.
> The *sanjū,* beginning and end, *shin gyōsō.*[21]
> Beating time with a fan.
> The style of *iro.*
> The technique of ending a line or musical unit.
> The *ryo* and *ritsu* seven-tone scales.
> The five elements and yin-yang.
> The aspect of left and right, front and back of melody.
> The style of shadow, the style of light.
> The aspect of always remaining calm.
> The art of Jōruri is the creation of music for a text.

The above thirteen points are part of the oral tradition.

Takemoto Gidayū (1687 First Month)

Notes
Glossaries
Bibliography
Index

Notes

Abbreviation used in the notes: *ZJH—Zeami jūrokubushū hyōshaku.*

CHAPTER 1: INTRODUCTION

1. I use the term Jōruri for the pre-1800 puppet theater and Bunraku for the post-1800 puppet theater.
2. Zeami Motokiyo, "Saragaku dangi," *Zeami jūrokubushū hyōshaku,* ed. Nose Asaji, II (1944), 385.
3. "Kakyō," *ZJH,* I (1940), p. 322.
4. *ZJH,* I, p. 590.
5. "Ongyoku kowadashi kuden," *ZJH,* II, p. 8.
6. *ZJH,* II, pp. 277-278.
7. *ZJH,* II, p. 281.
8. See Konishi Jin'ichi, "Association and Progression: Principles of Integration in Anthologies and Sequences of Japanese Court Poetry, A.D. 900–1350," *Harvard Journal of Asiatic Studies* 21: 67-127 (1958).
9. *Kodai chūsei geijutsu ron,* ed., Hayashiya Tatsusaburō (1973), pp. 514–516.
10. Paul Claudel, "Bunraku" (1929), in *Oeuvres Complètes* 3 (1952), p. 228.
11. Claudel, "Lettre au Professeur Miyajima" (1926) in *Oeuvres Complètes* 3, p. 230. The German drama scholar Carl Hagemann (1871–1945) traveled to Japan on a world trip during 1914 where he recorded his impressions of Bunraku theater, which concur with Claudel's. "The puppets present even better theater than the human actors. Their acting is even stricter art, and in the highest intensity of expression, because of the absence of human encumbrances, it is entirely centered on the symbolic. Here the puppets act like humans and in the place of humans. All reality has vanished; all that remains is what alone counts: the metaphor, the image. The human, all-too-human element has been transposed into a new dimension and lifted to unconditional aestheticism. The separation of sound and gesture on the theatrical stage has been carried to its final consequence: singers and speakers provide the foundation; they sing and speak, while the puppets act, directed and moved by different men, who are artists like the others, entirely

following the meaning of the sung or spoken words, their descriptions and eruptions, making the meaning more condensed and more profound." Translated by Hans E. Pringsheim in *KBS Bulletin on Japanese Culture* 88: 11 (February–March, 1968). Other French writers, namely Jean-Louis Barrault and, more recently, Roland Barthes, have been fascinated by the "discontinuous" nature of the Bunraku puppet theater.

11. Claudel, "Lettre au Professeur Miyajima" (1926) in *Oeuvres Complètes* 3, p. 230.
12. Donald Keene, *The Battles of Coxinga* (1951), p. 94.
13. Claudel, "Bunraku," p. 229.
14. T. S. Eliot, *The Use of Poetry and the Use of Criticism* (1933), p. 153.
15. Wayne C. Booth, *The Rhetoric of Fiction* (1961), pp. 60–64.

CHAPTER 2: MUSICAL CONVENTIONS

1. Yūda Yoshio, "Ko-jōruri no fushishō," *Jōruri shi ronkō* (1975), p. 47.
2. Marshall McLuhan, *The Gutenberg Galaxy* (1962), p. 142.
3. Ibid., p. 200.
4. Yokoyama Tadashi, *Jōruri ayatsuri shibai no kenkyū* (1964), p. 6.
5. Ibid., p. 7.
6. *Ningyō jōruri* (1975), ed., Geinōshi kenkyūkai, (Nihon shomin bunka shiryō shūsei series, vol. 7), p. 7.
7. Yūda, "Chikamatsu jōruri shichigyōbon no kenkyū," *Jōruri shi ronkō,* pp. 208–232.
8. Yokoyama, p. 163.
9. Yūda, "Takemoto Chikugonojō no kōkeisha mondai ni tsuite," *Jōruri shi ronkō,* pp. 384–407.
10. Reprinted in *Kinsei geidō ron,* ed., Nishiyama Matsunosuke et al., Nihon shisō taikei series, vol. 61 (1972).
11. Shōkōsai Hanbei, *Shibai gakuya zue,* ed. Hattori Yukio, Kabuki no bunken series, Vol. 5, pp. 59–60. Chikaishi Yasuaki also quotes it in *Ayatsuri jōruri no kenkyū: zokuhen* (1965), pp. 39–40.
12. Chikaishi, ibid., pp. 43–44. The preface is found in *Ningyō jōruri,* p. 637.
13. Keene, *The Battles of Coxinga,* p. 94.
14. Reprinted in *Kinsei geidō ron.*
15. Charles J. Dunn and Torigoe Bunzō, *The Actors' Analects* (*Yakusha Rongo*), (1969), pp. 78–79.
16. *Kodai chūsei geijutsu ron,* pp. 800–801. The introduction is by Nakamura Yasuo.
17. *Ningyō jōruri,* p. 135.
18. *Hachijō kadensho,* reprinted in *Kodai chūsei geijutsu ron,* pp. 540–545.
19. For instance, Chikamatsu Hanji (1725–1783) quotes Kaganojō in his "Ongyoku shugyōden" (1773), *Ningyō jōruri,* pp. 211–213.
20. Ueda Kazutoshi and Higuchi Yoshichiyo, eds., *Chikamatsu goi* (1930), pp. 395–428.
21. Yūda, *Jōruri shi ronkō,* pp. 47–55.
22. Tsunoda Ichirō, *Ningyō-geki no seiritsu ni kansuru kenkyū* (1963), pp. 785–818.
23. *Ningyō jōruri,* p. 149.
24. Yokoyama, pp. 13, 50, 93, 94. See also *Chikamatsu jōruri-bon okugaki shūsei,* edited by Osakafu Bungei Konwakai (1960), pp. 23–29.
25. *Kinsei geidō ron,* p. 407.
26. "Shikadō," *ZJH,* II, 468. Translated in *Sources of Japanese Tradition,* ed., Wm. Theodore de Bary et al. (1964 paperback), vol. I, p. 297.
27. Eliot, "Tradition and Individual Talent," *Selected Essays* (1932), pp. 4–5.

28. *Geinō kiroku* I (1977), ed., Geinōshi kenkyūkai (Nihon shomin bunka shiryō shūsei series, Vol. 12). His personal name was Naonori (1642–1695). He was born in Echizen (Ono) and was the daimyo of Murakami, Himeji, Yamagata, and Shirakawa during his lifetime. Another important diary covering the years 1684–1717, *Ōmu rōchū ki* by Asahi Shigeaki, also gives us information on the makeup of a day of theater.
29. *Ningyō jōruri,* p. 28.
30. Chikamatsu's early plays such as *Shusse Kagekiyo* (1685) are much shorter than his plays after *Yōmei tennō shokunin no kagami* (1706). Japanese scholars have determined that chanters performed one whole act of the shorter earlier plays, but more than one chanter performed each act of the longer plays after around 1706. See Mori Shū, *Chikamatsu Monzaemon* (1959), pp. 67–69 and Yūda, "Kinsei jōruri no keisei," *Jōruri shi ronkō,* pp. 32–33.
31. *Chikamatsu jōruri shū* II, ed., Shuzui Kenji and Ōkubo Tadakuni (1959).
32. Keene, *The Battles of Coxinga,* p. 95.
33. Makoto Ueda, *Literary and Art Theories of Japan,* pp. 192–193.
34. Yūda, "Chikamatsu jōruri no kaishaku," *Jōruri shi ronkō,* p. 100.
35. Ueda, p. 194.

CHAPTER 3: MOSAIC FORM

1. "Ongyoku no butai kara mita Chikamatsu," *Jōruri shi ronkō,* pp. 108–109.
2. The word *ji* is also used to describe the third-person (often in verse) narrative portions of plays, but *ji* in this book refers to the style of delivery, whether it is a first- or third-person passage. For a musicologist's view of these notation terms in contemporary usage, see William Malm, "A Musical Approach to the Study of Japanese Jōruri," in Brandon, ed., *Chūshingura: Studies in Kabuki and the Puppet Theater* (1982).
3. "Chikamatsu no ongaku to kōsei," *Jōruri shi ronkō,* pp. 73–74. *Yōkyoku shū,* I (1960), p. 13.
4. *Ki no Kaion zenshū,* I (1977), 162.
5. There is confusion about how far the domain of each notation symbol extends, particularly those for pitch. The notation has been recorded as in the original.
6. Chikaishi, *Ayatsuri jōruri no kenkyū: zokuhen,* pp. 97–98.
7. *Sonezaki shinjū, Meidō no hikyaku, ta gohen,* ed., Yūda Yoshio, (1977).
8. Yūda, "*Horikawa nami no tsuzumi* shiken," *Jōruri shi ronkō*, pp. 192–197. Yūda gives a detailed analysis of Chikamatsu's technique in the first scene of this play.
9. Keene, *Major Plays of Chikamatsu* (1961), pp. 279–280.
10. Ibid., pp. 174–175.

CHAPTER 4: CYCLICAL IMAGINATION

1. Joseph Campbell in *The Hero with a Thousand Faces* (1949) describes the Western archetypal journey of the hero as a circle. A summary of this journey is given on pp. 245–246.
2. A. K. Reischauer, "Genshin's *Ojo Yoshu:* Collected Essays on Birth into Paradise," *TASJ,* 2d series 8: 16–97 (December 1930). Allan A. Andrews, *The Teachings Essential for Rebirth: A Study of Genshin's "Ōjōyōshū"* (1973), p. 110. Pure Land Buddhism, which developed out of the Tendai sect, tended to move away from the idea that the believer had to make the journey and towards the idea that the chasm could be crossed by faith in Amida's vow. The Tendai monk Genshin's (947–1017) *Ōjōyōshū* (Essentials for salvation) contains a journey through the various Buddhist realms and has been compared to Dante's *Divine Comedy.*

3. Carmen Blacker, *The Catalpa Bow: A Study of Shamanistic Practices in Japan* (1978).
4. Ibid., pp. 218–234.
5. Ibid., p. 233.
6. Kenneth Butler, "The Textual Evolution of the *Heike monogatari,*" *HJAS* 26: 5–51 (1966).
7. Keene, ed., *Twenty Plays of the Nō Theatre* (1970), p. 293.
8. Kyoko Motomochi Nakamura, tr., *Miraculous Stories from the Japanese Buddhist Tradition: The "Nihon ryōiki" of the Monk Kyōkai* (1973). The translator discusses the symbolism of visits to the other world in her introduction, pp. 49–60.
9. Gunji Masakatsu, *Kabuki no bigaku* (1975), pp. 131–153.
10. Gunji Masakatsu, *Kabuki no hassō* (revised 1978), p. 197.
11. Ibid., pp. 214–215.
12. *Geinō kiroku* I (1977), p. 483.
13. Wing-tsit Chan, tr., *A Source Book of Chinese Philosophy* (1963), p. 245.
14. "There are two great structures underlying poetic imagery, the cyclical and the dialectic. The cycle of nature, running through the phases of the day, the year, the circulation of water, the generations of human life, and the like, stretches like a backbone across the whole of literature. The separation of images into the contrasting worlds or states of mind that Blake calls innocence and experience, and that religions call heaven and hell, is the dialectical framework of literature, and is the aspect of it that enables literature, without moralizing to create a moral reality in imaginative experience." Northrop Frye, *The Stubborn Structure: Essays on Criticism and Society* (1970), p. 101.
15. Johannes Barth, "Kagekiyo. Eine Betrachtung zum japanischen historischen Schauspiel," Deutschen Gesellschaft für Natur-und Volkerkunde Ostasiens, *Jubiläumsband* I (1933), pp. 299–329.
16. *Jōruri keiko buri* (1777) refers to current preference for dramatic scenes among amateur chanters; reprinted in *Edo jidai bungei shiryō* I (1916). Also in volume 7 of *Sharebon taisei* (Collection of *sharebon*), edited by Mizuno Minoru et al. (Tokyo, Chūō Kōron Sha, 1980).
17. Chikaishi in *Ayatsuri jōruri no kenkyū: zokuhen* (1965) uses this phrase throughout the book as the central theme.
18. Hirosue Tamotsu, *Zōho Chikamatsu josetsu: kinsei higeki no kenkyū* (1957). See also an English translation of his article on Kagekiyo in *Concerned Theatre Japan* 1.1:34–48 (Spring 1970).
19. Ibid., pp. 7–8.
20. Ibid., p. 19.
21. Ibid., p. 30.
22. Ibid., p. 31.
23. Shinoda Jun'ichi, "*Shusse Kagekiyo* no seiritsu ni tsuite," *Kokugo kokubungaku* (June 1959), rpt. in *Chikamatsu* (Nihon bungaku kenkyū shiryō sōsho series, 1977), pp. 154–168.
24. Ibid., p. 159.
25. Hirosue, *Chikamatsu josetsu,* p. 30.
26. Shinoda, p. 160.
27. Kuroki Kanzō, *Kinsei engeki kōsetsu* (1929), pp. 257–261.
28. Translated by Charles Dunn in *The Early Japanese Puppet Drama* (1966).
29. Susan Matisoff, *The Legend of Semimaru: Blind Musician of Japan* (1978), p. 148.
30. Ibid., p. 146.
31. Ibid., p. 148.
32. Ibid., p. 232.

33. The two scenes not marked are on p. 239 beginning with the new paragraph 4 lines from the bottom of the page, and on p. 241 beginning with Semimaru's speech in the middle of the page.
34. Ibid., p. 241.
35. Ibid., p. 243.
36. Ibid., p. 243.
37. Ibid., p. 244.
38. Ibid., p. 245.
39. Ibid., p. 245.
40. Ibid., p. 246.
41. Ibid., p. 247.
42. "Kinsei jōruri no keisei," *Jōruri shi ronkō,* pp. 26–49. There is a translation of this article in *Acta Asiatica* 28: 20–41 (1975).
43. Mori Shū, pp. 66–70.
44. A translation by Donald Keene is available in: *The Battles of Coxinga,* 1951 (which also contains much historical information about the play), *Major Plays of Chikamatsu,* 1961, and a shorter paperback edition, *Four Major Plays of Chikamatsu.*
45. Keene, *Major Plays of Chikamatsu,* p. 196.
46. The unit goes to the last line on p. 196, in Keene, ibid.
47. Ibid., p. 253.
48. Ibid., p. 224.
49. Ibid., p. 253.
50. Ibid., p. 229, n. 65.
51. Hozumi Ikan, *Naniwa miyage,* ed., Ueda Kazutoshi (1904), p. 158.
52. Keene, *Major Plays of Chikamatsu,* p. 252.
53. Ibid., pp. 252–253.
54. Ibid., p. 257.
55. Ibid., pp. 257–258.

CHAPTER 5: DESCENT TO PARADISE

1. *Ningyo jōruri,* ed., Geinō shi kenkyūkai, (Nihon shomin bunka shiryō shūsei series, vol. 7), p. 10.
2. Kuroki Kanzō, *Jōruri shi* (1943), p. 329, and Chikaishi, *Ayatsuri jōruri no kenkyū: zokuhen,* p. 170.
3. Keene, *World Within Walls: Japanese Literature of the Pre-Modern Era, 1600–1867* (1976), p. 256.
4. "*Sonezaki shinjū* michiyuki no kōsei," in Yūda, *Jōruri shi ronkō,* pp. 121-131.
5. *Sonezaki shinjū,* pp. 43–45.
6. This play is contained in Yūda's book (n. 5) and is divided into musical units according to the notation.
7. Ends with the line "meidō no mukai zo ya," *Chikamatsu Monzaemon shū* I, ed., Mori Shū et al. (1972), p. 186.
8. Ends with "yohan no shigure to narinikeri," *Chikamatsu Monzaemon shū* I, p. 187.
9. Ends with "kyūsenben ni zo hayanarinu," *Chikamatsu Monzaemon shū* I, p. 188.
10. *Sonezaki shinjū,* pp. 88–90.
11. Yokoyama Tadashi, *Jōruri ayatsuri shibai no kenkyū,* p. 270.
12. The *michiyuki* of this play is analyzed in *Sonezaki shinjū,* pp. 177–179. There is a German translation of *Shinjū kasane izutsu* in Detlef Schauwecker, *Studien zu Chikamatsu Monzaemon, Zwei bürgerliche Puppenspiele, Sprachlicher Stil und Struktur, mit Kommentar (II),* 1975.

13. Gunji, *Kabuki no bigaku,* pp. 131–208.
14. Keene, *Major Plays of Chikamatsu,* p. 154.
15. Ibid., p. 142.
16. "Chikamatsu jōruri no kaishaku," *Jōruri shi ronkō,* pp. 84–103. This article was unfinished when he died.
17. Divisions are from Yūda, *Sonezaki shinjū*, pp. 277–279.
18. Donald H. Shively, *The Love Suicide at Amijima: (Shinjū Ten no Amijima): A Study of a Japanese Domestic Tragedy by Chikamatsu Monzaemon* (1953), p. 41.
19. Keene, *Major Plays of Chikamatsu,* p. 89.
20. Ibid., p. 395.
21. Yūda, *Zenkō "Shinjū ten no Amijima"* (1975).
22. Ibid., pp. 355–356.
23. *Kinsei geidō ron,* pp. 416–438.
24. Yūda, *Zenkō "Shinjū ten no Amijima,"* p. 291.
25. Keene, *Major Plays of Chikamatsu,* p. 410.
26. Keene, *Some Japanese Portraits* (1978), p. 115.
27. Yokoyama Tadashi, *"Shinjū futatsu hara-obi, Shinjū yoigōshin* no ruten," *Gakudai kokubun* 11 (February 1968), rpt. in *Chikamatsu* (Nihon bungaku kenkyū shiryō sōsho series, 1976).
28. Keene, *Major Plays of Chikamatsu,* p. 418.

CHAPTER 6: CIRCLES OF FELICITY

1. The following is a list of Chikamatsu's *sewamono* plays that fall into this category: *Satsuma uta* (Song of Satsuma, 1704), *Tamba Yosaku matsuyo no Komuro-bushi* (Yosaku from Tamba, 1708), *Yodogoi shusse taki nobori* (A carp's successful climb up the waterfall, 1708), *Yūgiri Awa no Naruto* (Yūgiri and Izaemon, 1712), *Nagamachi onna harakiri* (Woman's *harakiri* in Nagamachi, 1712), *Daikyōji mukashi goyomi* (The almanac maker, 1715), *Yamazaki Yojibei nebiki no kadomatsu* (The uprooted pine, 1718), and also one by Ki no Kaion: *Wankyū sue no Matsuyama* (Wankyū and Matsuyama, 1708). There is a German translation of *Yodogoi shusse taki nobori* in Detlef Schauwecker, *Studien zu Chikamatsu Monzaemon, Zwei bürgerliche Puppenspiele, Sprachlicher Stil und Strukter, mit Kommentar,* II (Kyoto, 1975).
2. The plays *Satsuma* and *Tamba* have samurai as the lead figures, and *Yodogoi, Yūgiri, Yamazaki,* and *Wankyū* all have wealthy townsmen as the male protagonists.
3. Frye, *The Secular Scripture: A Study of the Structure of Romance* (1976), p. 100.
4. Ibid., p. 100.
5. *Chikamatsu no kenkyū*, ed., Tsubouchi Shōyō et al. (1900), p. 483.
6. *Zen'yaku Chikamatsu kessaku shū* II, ed., Wakatsuki Yasuji (1928), p. 184.
7. *Chikamatsu jōruri shū* I, ed., Shigetomo Ki (Nihon koten bungaku taikei series, vol. 49, 1958), p. 92.
8. Hirosue, *Chikamatsu josetsu,* p. 299.
9. *Chikamatsu no kenkyū,* p. 483.
10. *Chikamatsu josetsu,* p. 315.
11. Kuroki Kanzō, *Kinsei Nihon geinō ki* (1943), pp. 131–149, and Suwa Haruo, *Chikamatsu sewa-jōruri no kenkyū* (1974), pp. 252–263.
12. Translation from Keene, *Major Plays of Chikamatsu,* p. 123.
13. Suwa, *Chikamatsu sewa-jōruri no kenkyū,* p. 254.
14. Suwa notes a *Kabuki hyōbanki* reference from the year 1744 about a 1677 Kabuki performance, *Chikamatsu sewa-jōruri no kenkyū*, p. 255.

15. Reprinted in *Genroku kabuki kessaku shū* II, eds., Takano Tatsuyuki and Kuroki Kanzō (1973).
16. Kuroki, *Kinsei Nihon geinō ki,* p. 137.
17. Ibid., p. 141.
18. *Meisaku kabuki zenshū*, IV, eds., Toita Yasuji et al. (1970), p. 38.
19. Keene, *Major Plays of Chikamatsu,* p. 123.
20. Yokoyama, *Jōruri ayatsuri shibai no kenkyū,* p. 274.
21. Yūda, ed., *Sonezaki shinjū,* pp. 230–231.
22. Keene, *Major Plays of Chikamatsu,* p. 126.
23. Ibid., p. 127.
24. Ibid., p. 130.
25. Ibid., p. 321.
26. Ibid., p. 350.
27. Ibid., pp. 339–340.

APPENDIX A: PREFACE TO *A COLLECTION OF BAMBOO SHOOTS*

1. *Takenokoshū.* The text used in preparing the translation is found in *Kinsei geidō ron,* eds., Nishiyama Matsunosuke et al. (1972). Gunji Masakatsu annotated this text. The collection to which this is an introduction contains individual acts or scenes from Kaganojō's own plays. From the nature of the preface it is obvious that the purpose of publication was to serve the needs of amateur or novice chanters. See *Kaganojō danmonoshū*, ed., Yokoyama Shigeru (1958), for the complete text. Several texts are extant so there are no textual problems. These two treatises in the appendixes are discussed in Chapter 2, "Musical Convention."
2. Fujiwara Teika (1162–1241) in his *Eiga taigai* (1221?) states that there are no teachers for *waka* poetry. He then states that one should, however, learn from earlier poetry. See *Karon shū,* edited by Hashimoto Fumio et al., (Nihon Koten Bungaku Zenshū series, vol. 50, 1975), p. 495.
3. Kaganojō is following the Nō tradition in tracing his art's origins to antiquity. Zeami, in *Fushi Kaden,* begins Nō with the dance of the god Ameno Uzume no Mikoto recorded in the *Kojiki* (712). See *Zeami Zenchiku,* edited by Omote Akira and Katō Shūichi (1974), p. 14. The later *Hachijō kadensho* echoes this in its opening lines. See *Kodai chūsei geijutsu ron* (1973), p. 512.
4. He follows the *Hachijō kadensho.* See *Kodai chūsei geijutsu ron,* pp. 540–545.
5. The poem appears in the *Hachijō kadensho* and other Genroku-period works. It was thought to be Zeami's poem. The poem revolves on the double meaning of *fushi:* bamboo node or melody. See *Kinsei geidō ron,* note on p. 562.
6. Kaganojō seems to want Jōruri to be like the refined and classical art of Nō. He quotes from the *Hachijō kadensho* (*Kodai chūsei geijutsu ron,* p. 545).
7. Chanters at the time were behind a screen out of view of the audience. (See Figures 7 and 8.)
8. It was common practice in the period to give outstanding performing artists such titles. Kaganojō was known as Kadayū before receiving the title in 1677.
9. These terms are explained in the text, but in simpler language *ji* means a (dramatically) sung part; *kotoba* is straight speech; *ji iro* is rhythmical chanting (parlando) between *ji* and *kotoba* and is accompanied by music; *fushi* means a (lyrically) sung part (for example, *michiyuki*).
10. This style of heightened dramatic delivery was an important development which became a distinguishing characteristic of Jōruri. As is evident from the explanation, the chanter does not follow the samisen's melody. This interplay between the samisen and the chanter is the essence of the Jōruri art even today.

11. This refers to the pitch of the chanter's voice (explained in terms of the samisen): *haru* high (third string); *u* lighter but lower (second string); *chū* lowest, serious (first string). *Kinsei geidō ron,* note, p. 562.
12. The next three notations refer to transitional melodies. *Suete* is a style of chanting that is often used at climactic points in sad scenes. It marks a transition between two styles of chanting and is a partial cadence. *Okuri* is a more important transitional cadence; a character may enter or leave, or there may be a scene change. *Sanjū* is a cadence which marks a change of scenery or at least a major transition between scenes in the play.
13. This is a reference to Izumi Shikibu's poem translated by Arthur Waley as:

Kuraki yori	Out of the dark
Kuraki michi ni zo	Into a dark path
Irinubeki	I must now enter
Haruka ni terase	Shine (on me) from afar
Yama no ha no tsuki	Moon of the mountain fringe!

Japanese Poetry: The 'Uta,' (London, Oxford University Press, 1919), p. 86.

APPENDIX B: PREFACE TO *THE 1687 GIDAYŪ COLLECTION OF JŌRURI SCENES*

1. *Jōkyō yonen Gidayū danmonoshū.* The text used in preparing the translation is in *Kinsei geidō ron.* The text Gunji annotated was a copy made by Noma Kōshin, which was dated but without a title. The Tokugawa-period original was lost during or before World War II. Noma told Gunji that he found the manuscript hard to decipher (*Kinsei geidō ron,* p. 407), I have divided the translation into more paragraphs than the Japanese text has in order to make it more easily readable in English. These two treatises in the appendixes are discussed in Chapter 2, "Musical Convention."
2. The poem alluded to is by Yamabe Akahito and is found in the *Man'yōshū* as well as the *Shinkokinshū.*

Momoshiki no	The courtiers
Omiyabito wa	Of the Hundred Stone Palace
Itoma are ya	Dwell in such leisure!
Sakura kazashite	Cherry blossoms in their hair—
Kyo mo kurashitsu	Today again is passed in play

3. From the preface to the *Kokinshū.* The idea is one of classical balance between the two elements. Fujiwara Teika in his *Maigetsusho* (1219) makes the statement that "flower" means word and "fruit" means heart. See *Karon shū,* edited by Hashimoto Fumio et al., (Nihon Bungaku Zenshū series, vol. 50, 1975), p. 519.
4. By this he means that in this lively age one must chant vigorously, even to the point of exaggerating the language of the play, to produce a pleasing effect. For a Kabuki sense of *hana* (flower) and *mi* (fruit), see Dunn, ed., *The Actors' Analects* (1969), pp. 56–57.
5. The *Kadensho* is the seventeenth-century *Hachijō kadensho* Gidayū is perhaps referring to Kaganojō in this passage.
6. Perhaps this passage means more literally "imitate skill (of essence); do not imitate skill (of performance)." See pp. 31–32 where I have discussed this sentence.
7. This is a reference to the secret teachings of *waka* poetry, *Kokin godenju,* handed down within the Imperial Court, *Kinsei geidō ron,* p. 409.
8. Gidayū also takes his general ideas from the seventeenth-century *Hachijō kadensho.* See *Kodai chūsei geijutsu ron,* pp. 514–516.
9. This passage is usually only a few lines long. In a later work by Gidayū, *Tōryū*

jōruri kohyakuban 1702 or 1703 (*Ningyō jōruri*, p. 136), he explains the three types of openings in more detail. He says that the opening must introduce the essential theme of the entire work. The three kinds are: phrases from ancient classical (Chinese or Buddhist) writings; quotations from classical poetry; and quotations from contemporary songs. Each opening sets a different mood for the play. His three categories in this early work (1687) refer to the type of opening delivery style rather than specifically to the content.

10. In general, acts are divided into sections: *kuchi,* introduction, (*naka,* middle) and *kiri,* finale, and the act builds in intensity generally on the model of *jo-ha-kyū*. The first part(s) of an act are introductory to the final *kiri* scene, which is always performed by the best chanter.
11. Shura is one of the Buddhist realms the soul travels through after death in battle, the realm of never-ending fighting. Gidayū uses the word to describe the atmosphere of a vigorous warrior in lively action.
12. The *kuse* is the climactic part of many Nō plays. Gidayū seems to use the word in a double sense: one, in the context of the structure of a single Nō play in which the *kuse* is the climax, and, two, in the context of the orchestration of a program of plays in which the intensely tragic plays in the middle are the climax of the day's program.
13. *Jo, ha,* and *kyū* all have *jo-ha-kyū* divisions.
14. The fifth act must end auspiciously. A later (1737) *danmonoshū, Michinoku soma no yamabiko,* by Michinoku Modayū (a disciple of Gidayū) states that in the fifth act good and evil must be distinguished and the finale must be chanted gaily. See *Ningyō jōruri,* pp. 163–164.
15. The following section consists of technical notes on the art of chanting. They were intended as reminders for those who had already received oral instruction. The play quoted from is *Satsuma no kami Tadanori* (1686); the lines are from the *michiyuki* (in *Chikamatsu zenshū,* II, ed., Fujii Otoo, pp. 442–443). The story is about the Taira warrior Tadanori who risked his life to give a poem to Fujiwara Shunzei so that it would be included in the *Senzaishū* (1187–1189) imperial anthology.
16. These ideas relating chanting to a woven garment come from the *Hachijō kadensho* (*Kodai chūsei geijutsu ron,* pp. 545–546). *Tate* means narrow or weak (voice); *nuki* means full or robust.
17. Here Gidayū takes a line from the play and gives three marks of notation. The first is *tate,* next *shin no kiri* (break in mood) followed by *nuki chijimu* (robust but compressed), and finally *tate* again. I have changed a word (*natsu,* summer, to *koko,* here) because both Gidayū's and Kaganojō's original texts of this play, as well as a later *danmonoshū* of Gidayū, *Takemoto hiden maru* (1705 or 1706), have *koko.* Gidayū divides the lines into four movements. The first *tate* is introductory. The scene is a general setting along the ocean at Suma Bay. The *shin no kiri* lines focus on the speaker of the lines as her thoughts are stirred by the familiar salt breeze. The *kiri* (break) sense is that the chanter must shift to the *nuki* compressed line in which the piercing agony of her sad thoughts is expressed. (Tadanori's wife, Kikunomae, is thinking of Tadanori's inevitable death on the shores of Suma Bay where she now travels.) The above line on following the samisen or going against it makes sense in reference to this passage, because the quotation is at the intense moment when the woman's body is pierced by the wind. (Later, dressed as a warrior, she tries to die in Tadanori's place.) Gidayū shows a keen sense of balance between the content and style in the voice. The movement in the passage could perhaps be likened to a wave as it builds and finally cracks on the rocky shore, dissipating into the sand.

18. Perhaps the same connection between content and style is implied here: high rising smoke (*kun,* ruler) as opposed to the eye along the lower sea (*shin,* subject).
19. *Hachijō kadensho,* in *Kodai chūsei geijutsu ron,* p. 546. The idea here is that a one-*shaku* measure is a metaphor for the melody. The division is between the heart and the voice; the voice chants all of the words, but the heart varies the emphasis behind the words.
20. *Kinsei geidō ron,* note, p. 413. The drummer must carefully strike the drum lightly after each syllable.
21. These are types of cadences for the end of the scenes. *Shin* (*no sanjū*) is used only at the end of the first scene of Act One. *Gyōsō* (*no sanjū*) are all other types of *sanjū.* See *Tōryū jōruri kohyakuban* in *Ningyō jōruri,* p. 137.

Glossary A
Theatrical Terms

For a more detailed and extensive glossary of terms used in Bunraku (particularly for plays after Chikamatsu), see Yūda Yoshio's *Bunraku jōruri shū* (Iwanami, 1965).

Age 上	See *ue*
Aishō 哀傷	Pathos, tragedy
Aradate 荒立	Puppeteers' first practice of roles for a new play, while playwright recites text.
Banzuke 番付	Playbill listing title and performers.
Bunraku 文楽	A term now used for *ningyō jōruri* (Jōruri puppet theater). Uemura Bunrakuken (1750–1810) is supposed to have come to Osaka from Awaji and set up a Jōruri theater in the last decade of the eighteenth century. In 1811 after his death a theater called Bunraku was founded by his successor. The theater had many difficulties until 1856 when a theater was erected on the grounds of an Inari shrine in Bakuromachi. This theater organization blossomed during the golden age of Bunraku in the Meiji period (1868–1912).
Bun'ya (bushi) 文弥(節)	Style of singing very sadly in the manner of Okamoto Bun'ya who was active in Jōruri in the second half of the 17th century.
Chariba ちゃり場	Conventional comic scene in Kabuki or Bunraku.

Chū (Naka) 中	Notation: refers to low (or falling) pitch of chanter. Often used as transition from *ji* or *ji iro* to *kotoba;* similar to *iro.*
Dan 段	Has various uses; essentially it refers to portion of play one chanter performs. But since the amount a chanter performed changed over the years it can refer both to an act of a five-act play and to a part of an act performed by one chanter.
Danmari 黙	Conventional scene in Kabuki or Jōruri which takes place in silence and darkness; that is, relative darkness because the audience can still see the action. Movement is stylized and dancelike.
Danmonoshū 段物集	Collections of popular acts and scenes from Jōruri plays published for amateur chanters.
Edo (bushi) 江戸(節)	Notation: Edo style of Jōruri brought into Osaka play. Forceful, vigorous style.
Eiri-jōruri (bon) 絵入浄瑠璃(本)	17th and early 18th century (sometimes abbreviated) texts of Jōruri plays including pictures and some musical notation.
Ezukushi jōruri (bon) 絵尽し浄瑠璃(本)	18th century picture-book versions of plays in which pictures are large and text minimal.
Furyū (odori) 風流(踊り)	Term for dance which became popular approximately from the 12th century; it is usually associated with festivals. It was a dance in which many people joined in a line, often forming circles. The participants sang and played instruments while dancing.
Fushi 節, フシ	General word for melody. As notation it is written in *katakana:* 1. used to denote musical cadence; 2. used in song parts of the play.
Fushigoto 節事	Song portion of play (*michiyuki,* etc.); contrasts with dramatic *jigoto* parts.
Ge 下	See *on*
Gidayū (bushi) 義太夫(節)	Name of dominant style of Jōruri chanting developed by Takemoto Gidayū (1651–1714). Also used as a general term for Jōruri.
Giri 義理	A difficult term to pin down. It refers to "proper" behavior and has been translated as duty or honor. It refers to "rational" behavior in contrast to "emotional" behavior (see *ninjō*). In

	Chikamatsu's plays often *giri,* determined by the rational norms of society, is in opposition to individual human feelings and passions. Therefore in Chikamatsu's drama the *giri-ninjō* opposition can be considered as one of reason vs. desire.
Ha 破	See *jo-ha-kyū.*
Hana 花	Flower. An aesthetic term used in various ways. Zeami uses it in Nō to describe the beauty or perfection of an actor's art. Gidayū uses the term as in the *waka* poetic tradition of Teika who says that flower is the words of a poem and *mi* (fruit) is the *kokoro* (heart or feelings). In Kabuki the contrast between *hana* and *mi* is seen as between nonrealistic (dance) and realistic acting.
Haru ハル	Notation: third highest pitch. Voice is taut, and words are sung slowly.
Higeki 悲劇	Late 19th century translation of Western concept of tragedy.
Hosshin 発心	Buddhist term: awakening. Used to describe the experience of a day of Nō theater.
Iro 色	Notation: chanting style in between speech and song. Often used as transition between two styles.
Ji 地 (also later called ji ai 地合)	Notation: melodic delivery style accompanied by samisen.
Jidaimono 時代物	Historical or period play (usually in 5 *dan*) in which the setting is most often before 1600.
Jigoto 地事	Dramatic parts of play (in contrast to *fushigoto* song portions).
Ji iro 地色	Notation: rhythmical delivery style (parlando); samisen plays as well but not to harmonize with, rather to run parallel to, the chanter's voice.
Jo 序	See *jo-ha-kyū.*
Jo chū ge 上.中.下	Terms used to denote three *maki* of *sewamono* play.
Jo-ha-kyū 序破急.	Concept of musical structure: introduction, intensification, rapid conclusion. Usually divided

	into five parts: *jo*, ha_1, ha_2, ha_3, *kyū*. Terminology originated with court music and dance introduced from the continent in the 8th century.
Jōruri 浄瑠璃	Style of singing in puppet theater that developed from 1600; or a general term for the puppet theater.
Kabuki hyōbanki 歌舞伎評判記	Actor critiques usually published once a year in cities of Kyoto, Osaka, and Edo.
Kaeshi 返し	The practice of beginning with the final phrase of previous scene when new chanter begins his performance.
Kagura 神楽	Sacred music and dance. Also, the background music in Bunraku when a *jidaimono* scene takes place at or near a shrine.
Kakari カカリ	Notation: a high-pitched transition from *kotoba* to *ji* used in intensely sad or tragic scenes.
Kami 上	See *ue*
Kan 上	See *ue*
Kanazōshi 仮名草紙	17th century woodblock printed texts, mostly in *kana* syllabary.
Kin (Gin) キン	Notation: very high pitch (above *haru*) used by chanter at points of great emotion. Lively melody.
Kiri キリ（切リ）	The final part of act or scene where climax occurs. Also, the final part of 5 divisions of *michiyuki*.
Ko-jōruri 古浄瑠璃	Jōruri which predates Chikamatsu. *The Soga Heir* (1683) and *Kagekiyo Victorious* (1685) are considered transitional works between *ko-jōruri* and Jōruri.
Kotoba 詞	Notation: delivery style; speech in contrast to song (*ji*) and has no musical accompaniment.
Kōwaka-mai 幸若舞	Ballad drama, popular in the 15th and 16th centuries.
Ku 句	Line of poetry, song, etc. In abbreviated form, it is used as a punctuation mark in Jōruri texts to denote where the chanter takes a breath.
Kuchi 口	Opening part of act or scene and is introductory to final *kiri* where climax occurs.

Kudoki クドキ(口説き)	Conventional arialike style of expressing grief of characters. *Kudoki* passages are often the climax of scenes, acts, and whole plays.
Kuru クル	Notation: cadence which ends *kudoki.*
Kuse クセ	Notation used in Nō plays to denote a musical pattern; often found at climax of play. In Bunraku, it refers to a technique where the chanter alters the accent or length of a word to make it stand out for effect.
Kyū 急	See *jo-ha-kyū.*
Machibure taiko 町触太鼓	Parade of theater troupe through town to announce performance of a new play.
Mai 舞	A general term for dance, but used to contrast with *odori. Mai* is "classical" dance, that is, dance before the Tokugawa period (1600–1868).
Maki 巻	Term used to denote divisions of *sewamono* plays, which usually have 3 *maki.* The 5 *jidaimono* divisions are called *dan.* In Chikamatsu's time, 3 *maki* equal 1 *dan.*
Makura マクラ(枕)	Opening introductory portion of *michiyuki* or scene.
Maruhon 丸本 (Shōhon 正本)	"Official" Joruri texts published with chanter's acknowledgment that notation is correct.
Michiyuki 道行	Journey. Usually a pair travel from one place to another; it is a song and dance section of a play.
Mie 見得	Dramatic pose which is an important technique particularly in Kabuki but also used in Bunraku.
Migawari 身代り	The act of sacrificing oneself (or someone) for another. Often the crux of tragedy in Bunraku.
Mumyō 無明	Buddhist term for "ignorance" or "delusion." Literally, "absolute darkness."
Nai-keiko 内稽古	Private practice of new play.
Nai-yomi 内読	First reading by the playwright of a new play for theater troupe.
Naka 中	See *chū*
Ningyō-koshirae 人形拵	Preparation of puppet costumes.

Ninjō 人情	Literally, human feeling, sentiment, or desire; used in contrast to *giri*. Chikamatsu uses word *nasake* 情.
Noru ノル	Notation: delivery style in which voice "rides" lively rhythm of samisen.
Odori 踊り	General term for dance but used in contrast to *mai* to mean post-1600 dance. Also implies a more lively dance with leaps and jumps in contrast to classical *mai*.
Oiesōdō お家騒動	Kabuki plays in which there is a "disturbance in an important house." The disturbance usually concerns the successor of the head of a samurai or rich merchant household. The play ends with a happy solution.
Okina 翁	Ritual god play which begins a formal program of Nō. Also, general term for ritual dances that begin a program of Bunraku or Kabuki.
Okuri ヲクリ	Notation: cadence ending used at points when a character enters or leaves the stage. Also used when there is a change of chanters in performance.
On 下 (Ge, Sage, Shita)	Notation: lowest pitch used at moments of loneliness, quiet, anxiety, or fear.
Oroshi ヲロシ	Notation: cadence which marks end of formal prelude which opens *jidaimono* play.
Rankyoku 乱曲	Quick tempo, musical style.
Reizen 冷泉(節) (bushi)	Style of delivery used in lyrical passages; elegant delicate style.
Renbo 恋慕	Love.
Sage 下	See *on*
Sambasō 三番叟	Third dance of Okina-Senzai-Sambasō sequence used at opening of formal Nō performances. A lively dance which is also used to begin a Bunraku performance auspiciously.
Sanjū 三重	Notation: full musical cadence which ends scene in *jidaimono,* and *maki* in *sewamono.*
Sawari サワリ	Notation: used to mean the insertion of music other than Gidayū into play. Also, is used to mean *kudoki.*

Sekkyō 説教(経)	A general term for preaching on religious matters. Also used for performing arts aimed at teaching religious principles to the populace, as in *sekkyō jōruri.*
Senzai 千歳	Dance which follows Okina; auspicious.
Serifu 台詞、セリフ	Spoken parts in drama without musical accompaniment.
Sewamono 世話物	Plays of contemporary life (Tokugawa period 1600–1868) as opposed to *jidaimono*. Characters are usually not of upper classes. (Usually in 3 *maki.*)
Shinjū 心中	Term for love suicide in Jōruri plays. Two hearts become one in death.
Shita 下	See *on*
Shite シテ(仕手)	Protagonist in Nō plays.
Shōhon 正本	See *maruhon*
Shūgen 祝言	Auspicious.
Shura 修羅	Buddhist realm where one may go after dying in battle. Also used to describe state of mind or characteristics of personality associated with pride, anger, violence. In Jōruri, Shura refers to warriors in battle.
Sō-hon'yomi 惣本読	Reading of a new play by playwright before troupe members and everyone associated with the theater.
Sō-keiko 惣稽古	Public dress rehearsal.
Sō-odori 惣踊り	Final dance in Kabuki or Jōruri where whole cast joins in.
Suete スエテ	Notation: Semicadence used at moments of intense emotion. Samisen and voice produce harsh sounds to express feeling of character.
Sui 粋	Tokugawa-period term for *savoir faire,* elegance, taste, sensitivity, and sophistication. It is opposed to violence, money, and power.
Su-jōruri 素浄瑠璃	Chanting Jōruri without use of puppets. Gidayū says term means chanting without samisen.
Tan'i 単位	Unit, one part of a whole.
Tayū 太(大)夫	Bunraku chanter. Also term for highest class of

	courtesans, as well as general suffix used in names of performers.
Tsuzuki-kyōgen 続き狂言	Term for multiple-act Kabuki plays which developed in mid- to late-17th century.
U ウ	Notation: pitch above *chū*. Singing style is less taut, less melodic, and faster than *haru*.
Ue (Age, Kami, Kan) 上	Notation: highest pitch used at emotional moments.
Uki (Uku) 丂	Notation: high pitch above *u*. In between *haru* and *u;* also termed *haru u*.
Urei 愁	Pathos.
Uta 歌	Song. Notation: the insertion of a song into play.
Utai 謡い、ウタイ	Nō chanting. Notation in Jōruri: solemn style of chanting in the Nō manner.
Utazaimon 歌祭文	Ballad. Notation: ballad-style of singing.
Yamabushi 山伏	Priests of Shugendō sect whose beliefs are a mixture of both Buddhist and folk (shamanistic) concepts. They are considered to have superhuman powers, which they gain from ascetic pilgrimages to mountain shrines.
Yatsushi やつし(俏)	In Kabuki, a character who disguises himself as someone at the lowest level of the social scale. Also a term for plays on this theme. Often used in *oiesōdō* plays.
Yosegi zaiku 寄せ木細工	"Mosaic"—term used to describe musical structure of Nō and Jōruri because plays are composed of musical units of various sizes.
Yūgen 幽玄	Aesthetic term for elegance; for beauty which is mysterious and only partly seen.
Yuri ユリ(ユル) (Yuru)	Notation: chanter's technique of varying intonation during an emotional passage; individual syllables are lengthened in song.

Glossary B
Major Musical Notation

		MUSICAL CADENCE
Sanjū	三重	1) cadence indicates setting change. 2) used at the climax of battle scenes.
Okuri	ヲクリ	cadence indicates exit or entrance of a character and usually a change in the atmosphere on the stage.
Fushi	フシ	1) cadence which signals transition or end of primary unit. 2) style of singing of the *fushigoto* (song) sections of the play, such as *michiyuki*.
Suete	スエテ	musical pattern that expresses an intense emotion, usually sadness. It is a light cadence and often marks a division within a primary unit.
		DELIVERY STYLES
Ji	地	sung with samisen accompaniment (if follows *sanjū*, *okuri*, or *fushi*, then it begins a new primary unit).
Ji iro	地色	sung, but less melodious than *ji;* begins new primary unit if it follows cadence.
Iro	色	1) transition between chanting styles. 2) delivery style between *ji iro* and *kotoba*.

Kotoba	詞	speech. No melody or musical accompaniment.
Haru fushi	ハルフシ	style used in songs (*fushigoto*). Often begins *fushigoto* units.
Noru	ノル	singing style in which voice rides lively rhythm of samisen.

PITCH OF CHANTER'S VOICE IN DESCENDING ORDER
(also refers to voice technique)

Ue (or Kan, Age, Kami)	上	highest pitch, emotional.
Kin (Gin)	キン	high pitch, lively melody.
Haru	ハル	high pitch, voice is taut, and words are sung slowly.
Uki	ウ̄	in between *haru* and *u;* also termed *haru u.*
U	ウ	lower pitch, singing style is less taut, less melodic, and proceeds more quickly.
Chū (or Naka)	中	falling pitch, often used as transition from *ji* (or *ji iro*) to *kotoba;* similar to *iro.*
On (or Shita, Ge, Sage)	下	lowest pitch.

Glossary C
Structural Units of Jōruri Plays

Line: indicated by ⊿ punctuation mark which is an abbreviation for *ku* 句, line, or, in this case, breath or pause unit.

Primary Unit: (This is my term for the Japanese word *tan'i* 単位.) two kinds of primary units: *jigoto* 地事 (dramatic) and *fushigoto* 節事 (musical-lyrical). The *jigoto* primary units begin with musical notation *ji* 地 or *ji iro* 地色 (sung passage) and end with notation for the conclusive musical cadence *fushi* フシ. The unit is generally based on the *jo* 序 *ha* 破 *kyū* 急 format, meaning that the tension in the unit builds in intensity and tempo to a peak, which is then followed by a relaxing cadence *fushi* or by a cadence signaling the end of a scene or act. The *fushigoto* primary unit begins with *haru fushi* ハルフシ and ends with a conclusive cadence *fushi.* A scene may consist of five to twenty of these primary units.

Scene: indicated by *okuri* ヲクリ, a transitional cadence which brings a character onto or off the stage. These *okuri* mark divisions within the following unit.

Stage Setting Change: indicated by *sanjū* 三重, a final cadence which draws a major section of the play to a close. In a *jidaimono* (history) play the *sanjū* marks a scene in the act. In a three-*maki* (scene) *sewamono* (contemporary-life) play the *sanjū* usually marks the end of a *maki.* Normally there is a change of chanters and musicians after this notation.

Bibliography

BIBLIOGRAPHIC NOTE

Chikamatsu texts are in the following collections listed in the bibliography:

Chikamatsu jōruri shū, I, II 近松浄瑠璃集 (A collection of Chikamatsu's Jōruri plays)

Chikamatsu Monzaemon shū, I, II 近松門左衛門集 (A collection of Chikamatsu's plays)

Chikamatsu sewamono zenshū, 3 vols. 近松世話物全集 (A collection of Chikamatsu's contemporary-life plays)

Chikamatsu zenshū, 12 vols. 近松全集 (Complete works of Chikamatsu)

Shōhon Chikamatsu zenshū, 36 vols. 正本近松全集 (A complete collection of Chikamatsu's Jōruri original texts). Contains facsimiles of original texts.

Zeami treatises, including the *Nōsakusho* 能作書 (On composing Nō plays), are in:

Zeami jūrokubushū hyōshaku, 2 vols. 世阿彌十六部集評釋 (A collection of Zeami's sixteen treatises)

Zeami shū 世阿彌集 (A collection of Zeami's writings)

Zeami Zenchiku 世阿彌禅竹 (Zeami Zenchiku)

The following Jōruri treatises are in the collection *Ningyō jōruri* 人形浄瑠璃 (The Jōruri puppet theater) listed in the bibliography:

Chikuhō koji 竹豊故事 (A history of the Takemoto and Toyotake theaters)

Imamukashi ayatsuri nendaiki 今昔操年代記 (A chronology of the puppet theater)

Jōkyō yonen Gidayū danmonoshū 貞享四年義太夫段物集 (The 1687 Gidayū collection of Joruri scenes)
Jōruri kogiku maru 浄瑠璃小菊丸 (Jōruri chrysanthemums)
Tōryū jōruri kohyakuban 当流浄瑠璃小百番 (One hundred Jōruri favorites)
Michinoku soma no yamabiko 陸奥杣谽谺 (Echoes from Michinoku)
Ongyoku shugyōden 音曲修業伝 (On music training)
Ōmu ga soma 鸚鵡ヶ杣 (Parrot Mountain)
Shichikushū 紫竹集 (A collection of black bamboo)
Takenokoshū 竹子集 (A collection of bamboo shoots)
Takemoto hiden maru 竹本秘伝丸 (Secret teachings of the Takemoto tradition)

The following Jōruri treatises are in *Kinsei geidō ron* 近世芸道論 (A collection of Edo-period treatises on arts):

Jōkyō yonen Gidayū danmonoshū 貞享四年義太夫段物集 (The 1687 Gidayū collection of Jōruri scenes)
Ongyoku kudensho 音曲口伝書 (Oral teachings on music)
Takenokoshū 竹子集 (A collection of bamboo shoots)

Nō plays are in the following collections listed in the bibliography:

Yōkyoku shū, 2 vols. 謡曲集 (A collection of Nō plays)
Yōkyoku taikan, 7 vols. 謡曲大観 (A comprehensive collection of Nō plays)

Ko-jōruri plays are in the following source listed in the bibliography:

Ko-jōruri shōhon shū 古浄瑠璃正本集 (A collection of old Jōruri plays)

BIBLIOGRAPHY

Adachi, Barbara. *The Voices and Hands of Bunraku.* Tokyo, Kodansha International, 1978.

Amata (Kyota) kyakushokujō 許多脚色帖 (A vast collection of theatrical materials). Manuscript, 1827.

Amida no munewari 阿彌陀胸割 (Amida's riven breast). Woodblock, 1680s.

Andrews, Allan A. *The Teachings Essential for Rebirth: A Study of Genshin's "Ōjōyōshū."* Tokyo, Sophia University, 1973.

Araki, James T. *The Ballad Drama of Medieval Japan.* Berkeley and Los Angeles, University of California Press, 1964.

Asahi Shigeaki 朝日重章. *Ōmu rōchū ki* 鸚鵡籠中記 1684–1717 (Diary of a parrot in a cage). Nagoya sōsho zokuhen, vols. 9–12. Nagoya, Nagoya-shi Kyōiku iinkai, 1965–1968.

Barth, Johannes. "Kagekiyo. Eine Betrachtung zum japanischen historischen Schauspiel." Deutschen Gessellshcaft für Natur- und Volkerkunde Ostasiens. *Jubiläumsband* I (1933).

Barthes, Roland. *Image-Music-Text.* Translated by Stephan Heath. New York, Hill and Wang, 1977.

Blacker, Carmen. *The Catalpa Bow: A Study of Shamanistic Practices in Japan.* London, George Allen and Unwin, Ltd., 1975.

Booth, Wayne C. *The Rhetoric of Fiction.* Chicago, The University of Chicago Press, 1961.

Brandon, James R., trans. *Kabuki: Five Classic Plays.* Cambridge, Harvard University Press, 1975.

Bunraku jōruri shū 文楽浄瑠璃集 (A collection of Bunraku plays). Edited by Yūda Yoshio 祐田善雄. Nihon koten bungaku taikei series, vol. 99. Tokyo, Iwanami Shoten, 1965.

Butler, Kenneth Dean. "The Textual Evolution of the *Heike monogatari,*" *Harvard Journal of Asiatic Studies* 26: 5–51 (1966).

Campbell, Joseph. *The Hero with a Thousand Faces.* Princeton, Princeton University Press, 1949.

Chan, Wing-tsit, trans. *A Source Book in Chinese Philosophy.* Princeton, Princeton University Press, 1963.

Chikaishi Yasuaki 近石泰秋. *Ayatsuri jōruri no kenkyū: sono gikyoku kōsei ni tsuite* 操浄瑠璃の研究 — その戯曲構成について (A study of the puppet theater: The structure of plays). Tokyo, Kazama Shobō, 1961.

———. *Ayatsuri jōruri no kenkyū: zokuhen* 操浄瑠璃の研究続編 (A study of the puppet theater: Continuation). Tokyo, Kazama Shobō, 1965.

Chikamatsu 近松. Edited by Shuzui Kenji 守随憲治 et al. Zōho kokugo kokubungaku kenkyūshi taisei series, vol. 10. Tokyo, Sanseidō, 1977.

Chikamatsu 近松. Edited by Nihon bungaku kenkyū shiryō kankōkai 日本文学研究資料刊行会. Nihon bungaku kenkyū shiryō sōsho series. Tokyo, Yūseidō, 1976.

Chikamatsu jōruri-bon okugaki shūsei 近松浄瑠璃本奥書集成 (Collection of postscripts in Chikamatsu's Jōruri). Edited by Ōsaka-fu Bungei Konwakai 大阪府文芸懇話会 (Ōsaka Literary Salon). Osaka, Ōsaka Furitsu Toshokan, 1960.

Chikamatsu jōruri shū I 近松浄瑠璃集 上 (A collection of Chikamatsu Jōruri plays, vol. 1). Edited by Shigetomo Ki 重友毅. Tokyo, Iwanami Shoten, 1958.

Chikamatsu jōruri shū II 近松浄瑠璃集 下 (A collection of Chikamatsu Jōruri plays, vol. 2). Edited by Shuzui Kenji 守随憲治 and Ōkubo Tadakuni 大久保忠国. Nihon koten bungaku taikei series, vol. 50. Tokyo, Iwanami Shoten, 1959.

Chikamatsu kabuki kyōgen shū 近松歌舞伎狂言集 (A collection of Chikamatsu Kabuki plays). Edited by Takano Tatsuyuki 高野辰之. 2 vols. Tokyo, Rinsen Shoten, 1973. Rpt. of 1926 edition.

Chikamatsu Monzaemon shū I 近松門左衛門集 一 (A collection of Chikamatsu plays, vol. 1). Edited by Mori Shū 森修, Torigoe Bunzō 鳥越文蔵, and Nagatomo Chiyoji 長友千代治. Nihon koten bungaku zenshū series, vol. 43. Tokyo, Shōgakkan, 1972.

Chikamatsu Monzaemon shū II 近松門左衛門集 二 (A collection of Chikamatsu plays, vol. 2). Edited by Torigoe Bunzō 鳥越文蔵. Nihon koten bungaku zenshū series, vol. 44. Tokyo, Shōgakkan, 1975.

Chikamatsu no kenkyū 近松之研究 (A study of Chikamatsu). Edited by Tsubouchi Shōyō 坪内逍遥 et al. Tokyo, Shun'yōdō, 1900.

Chikamatsu sewamono zenshū 近松世話物全集 (A collection of Chikamatsu's contemporary-life plays). 3 vols. Edited by Fujii Otoo 藤井乙男. Tokyo, Fuzanbō, 1942.

Chikamatsu zenshū 近松全集 (Complete works of Chikamatsu). 12 vols. Edited by Fujii Otoo 藤井乙男. Osaka, Asahi Shimbun, 1923.

Chikuhō koji 竹豊故事 (A history of the Takemoto and Toyotake theaters). Woodblock, 1756.

Claudel, Paul. *Extrême Orient.* Vol. 3, in *Oeuvres Complètes.* Paris, Gallimard, 1952.

Crump, James I., and William P. Malm, eds. *Chinese and Japanese Music Dramas.* Michigan Papers in Chinese Studies, no. 19. Ann Arbor, Center for Chinese Studies, University of Michigan, 1975.

de Bary, Wm. Theodore, ed. *Sources of Japanese Tradition.* New York, Columbia University Press, 1958. (2 vol. paperback, 1964).

Dunn, Charles J. *The Early Japanese Puppet Drama.* London, Luzac and Company, Ltd., 1966.

——— and Torigoe Bunzō, ed. and trans. *The Actors' Analects (Yakusha Rongo).* New York, Columbia University Press, 1969.

Eliot, T. S. *The Use of Poetry and the Use of Criticism: Studies in the Relation of Criticism to Poetry in England.* London, Faber and Faber, Ltd., 1933.

———. *Selected Essays.* New Edition. New York, Harcourt, Brace and World, Inc., 1950.

Ezukushi kokusen'ya kassen 繪盡し國性爺合戰 (Illustrated battles of Coxinga). Woodblock, 1715.

Frye, Northrop. *The Stubborn Structure: Essays on Criticism and Society.* London, Methuen and Co., Ltd., 1970.

———. *The Secular Scripture: A Study of the Structure of Romance.* Cambridge, Harvard University Press, 1976.

Fujimoto Kizan 藤本箕山. *Shikidō ōkagami* 色道大鏡 (Great mirror of the way of love). Edited by Noma Kōshin 野間光辰. Kyoto, Yūzan Bunko, 1961.

Geinō kiroku I 芸能記録 一 (Records of performing arts, vol. 1). Edited by Geinōshi kenkyūkai 芸能史研究会. Nihon shomin bunka shiryō shūsei series, vol. 12. Tokyo, San'ichi Shobō, 1977.

Genroku kabuki kessaku shū 元禄歌舞伎傑作集 (A collection of Genroku Kabuki masterpieces). 2 vols. Edited by Takano Tatsuyuki 高野辰之 and Kuroki Kanzō 黒木勘蔵. 1925. Reprint. Tokyo, Rinsen Shoten, 1973.

Gidayū nenpyō: Kinseihen 義太夫年表：近世篇 (Chronology of Gidayū). 6 vols. Edited by Gidayū nempyō kinseihen kankōkai 義太夫年表近世篇刊行会. Tokyo, Yagi Shoten, 1979.

Gunji Masakatsu 郡司正勝. *Kabuki no bigaku* かぶきの美学 (Aesthetics of Kabuki). Tokyo, Engeki Shuppansha, 1975.

———. *Kabuki no hassō* かぶきの発想 (The origins of Kabuki). Tokyo, Meicho Kankōkai, revised edition, 1978.

Hachijō kadensho 八帖花伝書 (The eight-chapter *Kadensho*). In *Kodai chūsei geijitsu ron* 古代中世芸術論 (A collection of ancient treatises on arts). Edited by Hayashiya Tatsusaburō 林屋辰三郎. Nihon shisō taikei series, vol. 23. Tokyo, Iwanami Shoten, 1973.

Hara Michio 原道生. "'Yatsushi' no jōrurika–tabakouri Genshichi no mei to an" 「やつし」の浄瑠璃化 — 煙草売り源七の明と暗 ("Yatsushi" into Jōruri: The light and dark of "Tabako-uri Genshichi"), *Bungaku* 43: 627–650 (June 1975).

Hara Sumiko 原澄子. "Chikamatsu ni okeru kyokusetsu no mondai: Kaganojō to Gidayū no geifū o tōshite" 近松における曲節の問題 加賀掾と義太夫の芸風を通して (The question of musical notation in Chikamatsu's plays: With reference to Kaganojō's and Gidayū's styles), *Chikamatsu ronshū* (December 1964), pp. 25–37.

Hattori Yukio 服部幸雄. *Kabuki no kōzō: dentō engeki no sōzō seishin* 歌舞伎の構造 伝統演劇の創造精神 (The structure of Kabuki: The creative spirit of traditional theater). Tokyo, Chūō Kōron Sha, 1970.

———. *Hengeron: Kabuki no seishin shi* 変化論 歌舞伎の精神史 (Metamorphosis: A history of the spirit of Kabuki). Tokyo, Heibonsha, 1975.

Heike monogatari 平家物語 (The tale of the Heike). 2 vols. Edited by Ichiko Teiji 市古貞次. Nihon koten bungaku zenshū series, vols. 29 and 30. Tokyo, Shōgakkan, 1973.

Hirosue Tamotsu 広末保. *Zōho Chikamatsu josetsu: kinsei higeki no kenkyū* 増補近松序説近世悲劇の研究 (Introduction to Chikamatsu: A study of premodern tragedy [expanded version]). Tokyo, Miraisha, 1957.

———. "The Blind Kagekiyo: Beyond the Tragic?" *Concerned Theatre Japan* 1.1: 34–48 (Spring 1970).

Hoff, Frank, and Willi Flindt. "The Life Structure of Noh: An English Version of Yokomichi Mario's Analysis of the Structure of Noh." *Concerned Theatre Japan* 2. 3 and 4. Published in 1973 as a separate pamphlet.

Hori, Ichiro. *Folk Religion in Japan: Continuity and Change*. Edited by Joseph M. Kitagawa and Alan L. Miller. Chicago, The University of Chicago Press, 1968.

Hozumi Ikan 穂積以貫. *Naniwa miyage* 難波みやげ (Souvenir of Naniwa). Edited by Ueda Kazutoshi 上田萬年. Tokyo, Yūhōkan, 1904.

Imao Tetsuya 今尾哲也. *Henshin no shisō* 変身の思想 (A philosophy of transformation). Tokyo, Hōsei Daigaku Shuppan Kyoku, 1970.

Inobe Kiyoshi 井野辺潔. "Samisen to kyokusetsu" 三味線と曲節 (Samisen and musical notation) in *Jōruri* 浄瑠璃. Edited by Geinōshi kenkyūkai 芸能史研究会. Nihon no koten geinō series, vol. 7. Tokyo, Heibonsha, 1970.

Jōruri 浄瑠璃. Edited by Geinōshi kenkyūkai 芸能史研究会. Nihon no koten geinō series, vol. 7. Tokyo, Heibonsha, 1970.

Jōruri keiko buri 浄瑠璃稽古風流 (Jōruri lessons). In *Edo jidai bungei* I, *shiryō* 江戸時代文藝資料第一. 1777. Reprint, Tokyo, Kokusho Kankōkai, 1916.

Jōruri kenkyū bunken shūsei 浄瑠璃研究文獻集成 (A collection of Jōruri sources). Edited by Kawatake Shigetoshi 河竹繁俊. Nihon engeki bunken shūsei series, vol. 2. Tokyo, Hokkō Shobō, 1944.

Jōruri meisaku shū 浄瑠璃名作集 (A collection of famous Jōruri plays). 2 vols. Nihon meicho zenshū series, vol. 2. Tokyo, Nihon Meicho Zenshū Kankōkai, 1927.

Jōruri shū 浄瑠璃集 (A collection of Jōruri plays). Edited by Yokoyama Tadashi 横山正. Nihon koten bungaku zenshū series, vol. 45. Tokyo, Shōgakkan, 1971.

Kabuki hyōbanki shūsei 歌舞伎評判記集成 (A collection of Kabuki critiques). 11 vols. Edited by Kabuki hyōbanki kenkyūkai 歌舞伎評判記研究会. Tokyo, Iwanami Shoten, 1972.

Kaganojō danmonoshū 加賀掾段物集 (A collection of Kaganojō plays). Edited by Yokoyama Shigeru 横山重. Tokyo, Koten Bunko, 1958.

Karon shū 歌論集 (A collection of poetry treatises). Edited by Hashimoto Fumio 橋本不美男 et al. Nihon koten bungaku zenshū series, vol. 50. Tokyo, Shōgakkan, 1975.

Kawatake Shigetoshi 河竹繁俊. *Nihon engeki zenshi* 日本演劇全史 (A complete history of Japanese theater). Tokyo, Iwanami Shoten, 1959.

Keene, Donald. *The Battles of Coxinga: Chikamatsu's Puppet Play, Its Background and Importance.* London, Cambridge University Press, 1951.

———. *Japanese Literature: An Introduction for Westerners.* New York, Grove Press, 1955.

———. *Bunraku: The Art of the Japanese Puppet Theatre.* Tokyo, Kodansha International, 1965. Revised paperback, 1973.

———. *World Within Walls: Japanese Literature of the Pre-Modern Era, 1600–1867.* New York, Holt, Rinehart and Winston, 1976.

———. *Some Japanese Portraits.* Tokyo, Kodansha International, 1978.

———, trans. *Major Plays of Chikamatsu.* New York, Columbia University Press, 1961.

———, ed. *Twenty Plays of the Nō Theatre.* New York, Columbia University Press, 1970.

———, trans. *Chūshingura: The Treasury of Loyal Retainers.* New York, Columbia University Press, 1971.

Ki no Kaion zenshū 紀海音全集 (The complete works of Ki no Kaion). 8 vols. Osaka, Seibundō, 1977.

Kinpira jōruri shōhon shū 金平浄瑠璃正本集 (A collection of Kinpira Jōruri plays). 3 vols. Edited by Muroki Yatarō 室木弥太郎. Tokyo, Kadokawa Shoten, 1966.

Kinsei bungei no kenkyū: shiryōhen 近世文芸の研究: 資料篇 (A study of Edo-period arts: Sources). Edited by Shuzui Kenji 守随憲治. Tokyo, Kazama Shoin, 1974.

Kinsei geidō ron 近世芸道論 (A collection of Edo-period treatises on arts). Edited by Nishiyama Matsunosuke 西山松之助 et al. Nihon shisō taikei series, vol. 61. Tokyo, Iwanami Shoten, 1972.

Kitagawa, Hiroshi, and Bruce Tsuchida, trans. *The Tale of the Heike.* Tokyo, University of Tokyo Press, 1975.

Klopfenstein, Eduard. *Tausend Kirschbäume—Yoshitsune: Ein klassiches Stück des japanischen Theaters der Edozeit.* Studie, Übersetzung, Kommentar. Schweizer Asiatische Studien Monographien 2. Bern, Peter Lang, 1982.

Kodai chūsei geijutsu ron 古代中世芸術論 (A collection of ancient treatises on arts). Edited by Hayashiya Tatsusaburō 林屋辰三郎. Nihon shisō taikei series, vol. 23. Tokyo, Iwanami Shoten, 1973.

Ko-jōruri shōhon shū 古浄瑠璃正本集 (A collection of old Jōruri plays). 6 vols. Edited by Yokoyama Shigeru 横山重. Tokyo, Kadokawa Shoten, 1964.

Kokusen'ya kassen 国性爺合戦 (The battles of Coxinga). Woodblock reprint of 1715 text. Edited by Nihon koten bungakukai

日本古典文学会 . Tokyo, Nihon Koten Bungaku Kankōkai, 1973.

Konishi Jin'ichi. "Association and Progression: Principles of Integration in Anthologies and Sequences of Japanese Court Poetry, A.D. 900–1350," *Harvard Journal of Asiatic Studies* 21: 67–127 (1958).

Kurata Yoshihiro 倉田喜弘 . "Koten geinō kanshō nōto: Gidayū jōruri no keifu 1" 古典芸能鑑賞ノート 義太夫浄るりの系譜・1 (A note on appreciation of classical performing arts: The genealogy of Gidayū Jōruri, no. 1). *Kyōiku ongaku (chūgakuhan),* May 1975 , pp. 44–47.

———. "Koten geinō kanshō nōto: Gidayū jōruri no keifu 2" 古典芸能鑑賞ノート 義太夫浄るりの系譜・2 (A note on appreciation of classical performing arts: The Genealogy of Gidayū Jōruri, no. 2), *Kyōiku ongaku (chūgakuhan),* June 1975, pp. 34–37.

Kuroki Kanzō 黒木勘蔵 . *Kinsei engeki kōsetsu* 近世演劇考説 (A study of Edo-period theater). Tokyo, Rokkōkan, 1929.

———. *Jōruri shi* 浄瑠璃史 (A history of Jōruri). Tokyo, Seijisha, 1943.

———. *Kinsei Nihon geinō ki* 近世日本藝能記 (A study of Edo-period performing arts). Tokyo, Seijisha, 1943.

McLuhan, Marshall. *The Gutenberg Galaxy: The Making of Typographic Man.* London, Routledge and Kegan Paul, 1962.

Malm, William P. *Japanese Music and Musical Instruments.* Tokyo, Charles E. Tuttle Co., 1959.

———. "A Musical Approach to the Study of Japanese Jōruri," in *Chūshingura: Studies in Kabuki and the Puppet Theater.* Edited by James R. Brandon. Honolulu, University of Hawaii Press, 1982.

Matisoff, Susan. *The Legend of Semimaru: Blind Musician of Japan.* New York, Columbia University Press, 1978.

Matsudaira Naonori 松平直矩 . *Matsudaira Yamato no kami nikki* 松平大和守日記 (Diary of Matsudaira, governor of Yamato) in *Geinō kiroku* I 芸能記録 (Records of performing arts). Edited by Geinōshi kenkyūkai 芸能史研究会 . Nihon shomin bunka shiryō shūsei series, vol. 12. Tokyo, San'ichi Shobō, 1977.

Matsuzaki Hitoshi 松崎仁 . Genroku engeki kenkyū 元禄演劇研究 (A study of Genroku theater. Tokyo, Tōkyō Daigaku Shuppankai, 1979.

Meisaku kabuki zenshū 名作歌舞伎全集 (A complete collection of famous Kabuki plays). Vol. 4. Edited by Toita Yasuji 戸板康二 et al. Tokyo, Tōkyō Sōgen Shinsha, 1970.

Mitamura Engyo 三田村鳶魚 . "Yōkyoku to Uji Kaganojō" 謡曲と宇治加賀掾 (Nō and Uji Kaganojō), in *Mitamura Engyo zenshū* 三田村鳶魚全集 (Complete works of Mitamura Engyo). Vol. 12. Tokyo, Chūō Kōron Sha, 1976.

Mizutani Futō 水谷不倒 . *Eiri-jōruri shi* 絵入浄瑠璃史 (A history of illustrated Jōruri texts). *Mizutani Futō chosakushū* (Collected works of Mizutani Futō). Vol. 4. Tokyo, Chūō Kōron Sha, 1974.

Miyake Hitoshi 宮家準 . *Shugendō: yamabushi no rekishi to shisō* 修験道山伏の歴史と思想 (Shugendō: The history and thought of *yamabushi*). Tokyo, Kyōikusha, 1978.

Mori Shū 森修 . *Chikamatsu Monzaemon.* 近松門左衛門 (Chikamatsu Monzaemon). Koten to sono jidai series, vol. 4. Tokyo, San'ichi Shobō, 1959.

Mukai Yoshiki 向井芳樹 . *Chikamatsu no hōhō* 近松の方法 (Chikamatsu's method). Tokyo, Ōfūsha, 1976.

Nakamura, Kyoko Motomochi, trans. *Miraculous Stories from the Japanese Buddhist Tradition: The "Nihon ryōiki" of the Monk Kyōkai.* Cambridge, Harvard University Press, 1973.

Nakamura Yukihiko 中村幸彦 . *Kinsei bungei shichōkō* 近世文芸思潮攷 (Tokugawa literature and thought). Tokyo, Iwanami Shoten, 1975.

Ningyō jōruri 人形浄瑠璃 (The Jōruri puppet theater). Edited by Geinōshi kenkyūkai 芸能史研究会 . Nihon shomin bunka shiryō shūsei series, vol. 7. Tokyo, San'ichi Shobō, 1975.

Nō, kyōgen, bunraku 能狂言文楽 (Nō, Kyōgen, Bunraku). Kita Roppeita 喜多六平太 et al. Nihon no geidan series, vol. 3. Tokyo, Kyūgei Shuppan, 1978.

Ogakuzu 鋸末 (Sawdust). Woodblock, 1699.

Ōhashi Tadayoshi 大橋正叔 . "Kinsei jōruri geiron no mondai" (1) 近世浄瑠璃芸能論の問題（一） (Questions on Edo-period Jōruri treatises, 1), *Ōsaka shōin joshidai ronshū* 12: 33–46 (February 1975).

Origuchi Shinobu 折口信夫 . *Geinō shi hen* 芸能史篇 (History of the performing arts). 2 vols. *Origuchi Shinobu zenshū* (Complete works of Origuchi Shinobu), vols. 17 and 18. Tokyo, Chūō Kōron Sha, 1956.

———. *Nihon geinō shi* 日本芸能史 (A history of Japanese performing arts). *Origuchi Shinobu zenshū nōtohen* (Complete works of Origuchi Shinobu: Notes), vol. 5. Tokyo, Chūō Kōron Sha, 1972.

Reischauer, A. K. "Genshin's *Ojo Yoshu:* Collected Essays on Birth into Paradise," *Transactions of the Asiatic Society of Japan.* Second series 8: 16–97 (December 1930).

Ruri tengu 瑠璃天狗 1806. Edited by Kuroki Kanzō 黒木勘蔵 in *Ki no Kaion, Namiki Sōsuke jōrurishū* 紀海音・並木宗輔浄瑠璃集(Collection of Ki no Kaion and Namiki Sōsuke Jōruri). Teikoku Bunko. Tokyo, Hakubunkan, 1929.

Schauwecker, Detlef. *Studien zu Chikamatsu Monzaemon, Zwei bürger-*

liche Puppenspiele, Sprachlicher Stil und Struktur, mit Kommentar (II). Marburger Studien zur Afrika–und Asienkunde. Serie B: Asien Bd. 3. Kyoto, Selbstverlag, 1975.

Seikyoku ruisan 聲曲類纂 (A history of singing and music). Edited by Fujita Tokutarō 藤田德太郎. Tokyo, Iwanami Shoten, 1941.

Shinjū nimai ezōshi 心中二枚繪草紙 (Love suicides and double-folded picture books). Woodblock, 1706.

Shinoda Jun'ichi 信多純一. "*Shusse Kagekiyo* no seiritsu ni tsuite" 出世景清の成立について (On the background to *Kagekiyo Victorious*). *Kokugo kokubun* (June 1959). Reprint in *Chikamatsu* 近松. Edited by Nihon bungaku kenkyū shiryō kankōkai 日本文学研究資料刊行会. Nihon bungaku kenkyū shiryō sōsho series. Tokyo, Yūseidō, 1976.

Shively, Donald H. *The Love Suicide at Amijima (Shinjū Ten no Amijima): A Study of a Japanese Domestic Tragedy by Chikamatsu Monzaemon.* Cambridge, Harvard University Press, 1953.

Shōhon Chikamatsu zenshū 正本近松全集 (A complete collection of Chikamatsu's Jōruri original texts). 36 vols. Edited by Chikamatsu shoshi kenkyūkai 近松書誌研究会. Tokyo, Benseisha, 1977.

Shōkōsai Hanbei 松好斎半兵衛. *Shibai gakuya zue* 劇場楽屋図会 (Illustrations of backstage). 1802. Edited by Hattori Yukio 服部幸雄. Kabuki no bunken series, vol. 5. Tokyo, Kokuritsu Gekijō, 1974.

Sonezaki moyō 曾根崎模様 (Patterns of Sonezaki). Woodblock, 1761.

Sonezaki shinjū, Meido no hikyaku, ta gohen 曾根崎心中冥途の飛脚他五篇 (*Love Suicides at Sonezaki, Courier for Hell,* and five other plays). Edited by Yūda Yoshio 祐田善雄. Tokyo, Iwanami Shoten, 1977.

Suwa Haruo 諏訪春雄. *Genroku kabuki no kenkyū* 元禄歌舞伎の研究 (A study of Genroku Kabuki). Tokyo, Kazama Shoin, 1966.

———. *Chikamatsu sewa-jōruri no kenkyū* 近松世話浄瑠璃の研究 (A study of Chikamatsu's contemporary-life plays). Tokyo, Kazama Shoin, 1974.

Thornbury, Barbara. *Sukeroku's Double Identity: The Dramatic Structure of Edo Kabuki.* Michigan Papers in Japanese Studies, no. 6. Ann Arbor, Center for Japanese Studies, University of Michigan, 1982.

Torii Fumiko 鳥居フミ子. "Kaganojō jidai no Chikamatsu" 加賀掾時代の近松 (Chikamatsu's plays written for Kaganojō), *Chikamatsu ronshū,* 1 (September 1962). Reprint in *Chikamatsu* 近松 (Chikamatsu). Edited by Nihon bungaku kenkyū shiryō kankōkai 日本文学研究資料刊行

会. Nihon bungaku kenkyū shiryō sōsho series. Tokyo, Yūseidō, 1976.

Tōsei shibai katagi 當世芝居氣質 (Characteristics of modern theater). Woodblock, 1777.

Tsunoda Ichirō 角田一郎. *Ningyō-geki no seiritsu ni kansuru kenkyū* 人形劇の成立に関する研究. (A study of the development of the puppet theater). Osaka, Asahiya Shoten, 1963.

———. *Gidayū-bushi no keisei ni kansuru ichi-shiron (I)—Kazan-in no michiyuki ni tsuite* 義太夫節の形成に関する一試論 (一) 花山院の道行について (On the formation of Gidayū: The *michiyuki* of *Kazan-in*), *Kinsei bungei: kenkyū to hyōron* 5: 150–161 (November 1973).

———. "Ongakusei to geki kōsō: gen kenkyū dankai to kadai" 音楽性と劇構想・現研究段階と課題 (Music and dramatic structure: The state of research), *Kokubungaku kaishaku to kanshō* (September 1974), pp. 77–84.

Uchiyama Mikiko 内山美樹子. "Jōruri no gikyoku sahō" 浄瑠璃の戯曲作法 (The composition of Jōruri plays) in *Jōruri* 浄瑠璃. Edited by Geinōshi kenkyūkai 芸能史研究会. Nihon no koten geinō series, vol. 7. Tokyo, Heibonsha, 1970.

Ueda Kazutoshi 上田萬年, and Higuchi Yoshichiyo 樋口慶千代, eds. *Chikamatsu goi* 近松語彙 (Chikamatsu dictionary). Tokyo, Fuzanbō, 1930.

Ueda, Makoto. *Literary and Art Theories in Japan.* Cleveland, The Press of Western Reserve University, 1967.

Umehara Takeshi 梅原猛. *Jigoku no shisō: Nihon seishin no ikkeifu* 地獄の思想・日本精神の一系譜 (Philosophy of hell: A genealogy of the Japanese mind). Tokyo, Chūō Kōron Sha, 1967.

Urabe Kanji 浦部幹資. "Ningyō shibai gashō mokuroku kō" 人形芝居画証目録稿 (A list of puppet theater illustrations), *Engekigaku* 19:100–118 (March 1978).

Wakatsuki Yasuji 若月保治. *Chikamatsu ningyō-jōruri no kenkyū* 近松人形浄瑠璃の研究 (A study of Chikamatsu Jōruri). Tokyo, Daiichi Shobō, 1934.

———. *Ningyō-jōruri shi kenkyū: sanbyakunen shi kenkyū* 人形浄瑠璃史研究：三百年史研究 (Three hundred years of Jōruri). Tokyo, Sakurai Shoten, 1943.

Waley, Arthur. *The Nō Plays of Japan.* London, Allen and Unwin, 1921.

Watsuji Tetsurō 和辻哲郎. *"Nihon geijutsu shi kenkyū" kabuki to ayatsuri jōruri* 日本芸術史研究 歌舞伎と操浄瑠璃 (A study of Japanese arts: Kabuki and Jōruri). Tokyo, Iwanami Shoten, 1955.

Yasuda, Kenneth K. "The Structure of 'Hagoromo,' a Nō Play," *Harvard Journal of Asiatic Studies* 33: 5–89 (1973).

Yokoyama Tadashi 横山正. *Jōruri ayatsuri shibai no kenkyū* 浄瑠璃操芝居の研究 (A study of Jōruri puppet theater). Tokyo, Kazama Shoten, 1964.

———. *Kinsei engeki ronsō* 近世演劇論叢 (A collection of essays on Edo-period theater). Osaka, Seibundō, 1976.

———. "*Shinjū futatsu hara-obi, Shinjū yoigōshin* no ruten" 「心中二ツ腹帯」「心中宵庚申」の流転 (Performances of *Love Suicides of the Double Brocade Obi* and *Love Suicides on the Eve of the Kōshin Festival*), *Gakudai kokubun* 11 (February 1968). Reprint in *Chikamatsu* 近松. Edited by Nihon bungaku kenkyū shiryō kankōkai 日本文学研究資料刊行会. Nihon bungaku kenkyū shiryō sōsho series. Tokyo, Yūseidō, 1976.

Yōkyoku shū 謡曲集 (A collection of Nō plays). 2 vols. Edited by Yokomichi Mario 横道萬里雄 and Omote Akira 表章. Nihon koten bungaku taikei series, vols. 40 and 41. Tokyo, Iwanami Shoten, 1960.

Yōkyoku taikan 謡曲大観 (A comprehensive collection of Nō plays). 7 vols. Edited by Sanari Kentarō 佐成謙太郎. Tokyo, Meiji Shoin, 1927.

Yūda Yoshio 祐田善雄. *Jōruri shi ronkō* 浄瑠璃史論考 (Essays on the history of Jōruri). Tokyo, Chūō Kōron Sha, 1975.

———. *Zenkō "Shinjū ten no Amijima"* 全講心中天の網島 (A complete annotation of *Love Suicides at Amijima*). Tokyo, Shibundō, 1975.

———. "Ki no Kaion no chosaku nendai kōshō to sono sakuhin keikō (I)" 紀海音の著作年代考證とその作品傾向（上） (A chronology of the life and works of Ki no Kaion I), *Kokugo kokubun* 6: 41–93 (July 1936).

———. "Ki no Kaion no chosaku nendai kōshō to sono sakuhin keikō (II)" 紀海音の著作年代考證とその作品傾向（下） (A chronology of the life and works of Ki no Kaion II), *Kokugo kokubun* 6: 82–107 (August 1936).

———. "The Formation of Early Modern *Jōruri*," *Acta Asiatica, Bulletin of the Institute of Eastern Culture* 28: 20–41 (1975).

Zeami jūrokubushū hyōshaku 世阿彌十六部集評釋 (A collection of Zeami's sixteen treatises). 2 vols. Edited by Nose Asaji 能勢朝次. Tokyo, Iwanami Shoten, 1940, 1944.

Zeami shū 世阿彌集 (A collection of Zeami's writings). Edited by Konishi Jin'ichi 小西甚一. Nihon no shisō series, vol. 8. Tokyo, Chikuma Shobō, 1970.

Zeami Zenchiku 世阿彌禅竹 (Zeami Zenchiku). Edited by Omote Akira 表章 and Katō Shūichi 加藤周一. Nihon shisō taikei, vol. 24. Tokyo, Iwanami Shoten, 1974.

Zen'yaku Chikamatsu kessaku shū 全譯近松傑作集 (Complete translations of Chikamatsu's masterpieces). 3 vols. Edited by Wakatsuki Yasuji 若月保治. Tokyo, Taiyōdō Shoten, 1928.

Index

Harvard East Asian Monographs

1. Liang Fang-chung, *The Single-Whip Method of Taxation in China*
2. Harold C. Hinton, *The Grain Tribute System of China, 1845–1911*
3. Ellsworth C. Carlson, *The Kaiping Mines, 1877–1912*
4. Chao Kuo-chün, *Agrarian Policies of Mainland China: A Documentary Study, 1949–1956*
5. Edgar Snow, *Random Notes on Red China, 1936–1945*
6. Edwin George Beal, Jr., *The Origin of Likin, 1835–1864*
7. Chao Kuo-chün, *Economic Planning and Organization in Mainland China: A Documentary Study, 1949–1957*
8. John K. Fairbank, *Ch'ing Documents: An Introductory Syllabus*
9. Helen Yin and Yi-chang Yin, *Economic Statistics of Mainland China, 1949–1957*
10. Wolfgang Franke, *The Reform and Abolition of the Traditional Chinese Examination System*
11. Albert Feuerwerker and S. Cheng, *Chinese Communist Studies of Modern Chinese History*
12. C. John Stanley, *Late Ch'ing Finance: Hu Kuang-yung as an Innovator*
13. S. M. Meng, *The Tsungli Yamen: Its Organization and Functions*
14. Ssu-yü Teng, *Historiography of the Taiping Rebellion*
15. Chun-Jo Liu, *Controversies in Modern Chinese Intellectual History: An Analytic Bibliography of Periodical Articles, Mainly of the May Fourth and Post-May Fourth Era*
16. Edward J. M. Rhoads, *The Chinese Red Army, 1927–1963: An Annotated Bibliography*
17. Andrew J. Nathan, *A History of the China International Famine Relief Commission*
18. Frank H. H. King (ed.) and Prescott Clarke, *A Research Guide to China-Coast Newspapers, 1822–1911*
19. Ellis Joffe, *Party and Army: Professionalism and Political Control in the Chinese Officer Corps, 1949–1964*
20. Toshio G. Tsukahira, *Feudal Control in Tokugawa Japan: The Sankin Kōtai System*

21. Kwang-Ching Liu, ed., *American Missionaries in China: Papers from Harvard Seminars*
22. George Moseley, *A Sino-Soviet Cultural Frontier: The Ili Kazakh Autonomous Chou*
23. Carl F. Nathan, *Plague Prevention and Politics in Manchuria, 1910–1931*
24. Adrian Arthur Bennett, *John Fryer: The Introduction of Western Science and Technology into Nineteenth-Century China*
25. Donald J. Friedman, *The Road from Isolation: The Campaign of the American Committee for Non-Participation in Japanese Aggression, 1938–1941*
26. Edward Le Fevour, *Western Enterprise in Late Ch'ing China: A Selective Survey of Jardine, Matheson and Company's Operations, 1842–1895*
27. Charles Neuhauser, *Third World Politics: China and the Afro-Asian People's Solidarity Organization, 1957–1967*
28. Kungtu C. Sun, assisted by Ralph W. Huenemann, *The Economic Development of Manchuria in the First Half of the Twentieth Century*
29. Shahid Javed Burki, *A Study of Chinese Communes, 1965*
30. John Carter Vincent, *The Extraterritorial System in China: Final Phase*
31. Madeleine Chi, *China Diplomacy, 1914–1918*
32. Clifton Jackson Phillips, *Protestant America and the Pagan World: The First Half Century of the American Board of Commissioners for Foreign Missions, 1810–1860*
33. James Pusey, *Wu Han: Attacking the Present through the Past*
34. Ying-wan Cheng, *Postal Communication in China and Its Modernization, 1860–1896*
35. Tuvia Blumenthal, *Saving in Postwar Japan*
36. Peter Frost, *The Bakumatsu Currency Crisis*
37. Stephen C. Lockwood, *Augustine Heard and Company, 1858–1862*
38. Robert R. Campbell, *James Duncan Campbell: A Memoir by His Son*
39. Jerome Alan Cohen, ed., *The Dynamics of China's Foreign Relations*
40. V. V. Vishnyakova-Akimova, *Two Years in Revolutionary China, 1925–1927*, tr. Steven I. Levine
41. Meron Medzini, *French Policy in Japan during the Closing Years of the Tokugawa Regime*
42. *The Cultural Revolution in the Provinces*
43. Sidney A. Forsythe, *An American Missionary Community in China, 1895–1905*
44. Benjamin I. Schwartz, ed., *Reflections on the May Fourth Movement: A Symposium*
45. Ching Young Choe, *The Rule of the Taewŏn'gun, 1864–1873: Restoration in Yi Korea*

46. W. P. J. Hall, *A Bibliographical Guide to Japanese Research on the Chinese Economy, 1958–1970*
47. Jack J. Gerson, *Horatio Nelson Lay and Sino-British Relations, 1854–1864*
48. Paul Richard Bohr, *Famine and the Missionary: Timothy Richard as Relief Administrator and Advocate of National Reform*
49. Endymion Wilkinson, *The History of Imperial China: A Research Guide*
50. Britten Dean, *China and Great Britain: The Diplomacy of Commerical Relations, 1860–1864*
51. Ellsworth C. Carlson, *The Foochow Missionaries, 1847–1880*
52. Yeh-chien Wang, *An Estimate of the Land-Tax Collection in China, 1753 and 1908*
53. Richard M. Pfeffer, *Understanding Business Contracts in China, 1949–1963*
54. Han-sheng Chuan and Richard Kraus, *Mid-Ch'ing Rice Markets and Trade, An Essay in Price History*
55. Ranbir Vohra, *Lao She and the Chinese Revolution*
56. Liang-lin Hsiao, *China's Foreign Trade Statistics, 1864–1949*
57. Lee-hsia Hsu Ting, *Government Control of the Press in Modern China, 1900–1949*
58. Edward W. Wagner, *The Literati Purges: Political Conflict in Early Yi Korea*
59. Joungwon A. Kim, *Divided Korea: The Politics of Development, 1945–1972*
60. Noriko Kamachi, John K. Fairbank, and Chūzō Ichiko, *Japanese Studies of Modern China Since 1953: A Bibliographical Guide to Historical and Social-Science Research on the Nineteenth and Twentieth Centuries, Supplementary Volume for 1953–1969*
61. Donald A. Gibbs and Yun-chen Li, *A Bibliography of Studies and Translations of Modern Chinese Literature, 1918–1942*
62. Robert H. Silin, *Leadership and Values: The Organization of Large-Scale Taiwanese Enterprises*
63. David Pong, *A Critical Guide to the Kwangtung Provincial Archives Deposited at the Public Record Office of London*
64. Fred W. Drake, *China Charts the World: Hsu Chi-yü and His Geography of 1848*
65. William A. Brown and Urgunge Onon, translators and annotators, *History of the Mongolian People's Republic*
66. Edward L. Farmer, *Early Ming Government: The Evolution of Dual Capitals*
67. Ralph C. Croizier, *Koxinga and Chinese Nationalism: History, Myth, and the Hero*
68. William J. Tyler, tr., *The Psychological World of Natsumi Sōseki*, by Doi Takeo

69. Eric Widmer, *The Russian Ecclesiastical Mission in Peking during the Eighteenth Century*
70. Charlton M. Lewis, *Prologue to the Chinese Revolution: The Transformation of Ideas and Institutions in Hunan Province, 1891–1907*
71. Preston Torbert, *The Ch'ing Imperial Household Department: A Study of its Organization and Principal Functions, 1662–1796*
72. Paul A. Cohen and John E. Schrecker, eds., *Reform in Nineteenth-Century China*
73. Jon Sigurdson, *Rural Industrialization in China*
74. Kang Chao, *The Development of Cotton Textile Production in China*
75. Valentin Rabe, *The Home Base of American China Missions, 1880–1920*
76. Sarasin Viraphol, *Tribute and Profit: Sino-Siamese Trade, 1652–1853*
77. Ch'i-ch'ing Hsiao, *The Military Establishment of the Yuan Dynasty*
78. Meishi Tsai, *Contemporary Chinese Novels and Short Stories, 1949–1974: An Annotated Bibliography*
79. Wellington K. K. Chan, *Merchants, Mandarins, and Modern Enterprise in Late Ch'ing China*
80. Endymion Wilkinson, *Landlord and Labor in Late Imperial China: Case Studies from Shandong by Jing Su and Luo Lun*
81. Barry Keenan, *The Dewey Experiment in China: Educational Reform and Political Power in the Early Republic*
82. George A. Hayden, *Crime and Punishment in Medieval Chinese Drama: Three Judge Pao Plays*
83. Sang-Chul Suh, *Growth and Structural Changes in the Korean Economy, 1910–1940*
84. J. W. Dower, *Empire and Aftermath: Yoshida Shigeru and the Japanese Experience, 1878–1954*
85. Martin Collcutt, *Five Mountains: The Rinzai Zen Monastic Institution in Medieval Japan*

STUDIES IN THE MODERNIZATION OF THE REPUBLIC OF KOREA: 1945–1975

86. Kwang Suk Kim and Michael Roemer, *Growth and Structural Transformation*
87. Anne O. Krueger, *The Developmental Role of the Foreign Sector and Aid*
88. Edwin S. Mills and Byung-Nak Song, *Urbanization and Urban Problems*
89. Sung Hwan Ban, Pal Yong Moon, and Dwight H. Perkins, *Rural Development*

90. Noel F. McGinn, Donald R. Snodgrass, Yung Bong Kim, Shin-Bok Kim, and Quee-Young Kim, *Education and Development in Korea*

91. Leroy P. Jones and Il SaKong, *Government, Business and Entrepreneurship in Economic Development: The Korean Case*

92. Edward S. Mason, Dwight H. Perkins, Kwang Suk Kim, David C. Cole, Mahn Je Kim, et al., *The Economic and Social Modernization of the Republic of Korea*

93. Robert Repetto, Tai Hwan Kwon, Son-Ung Kim, Dae Young Kim, John E. Sloboda, and Peter J. Donaldson, *Economic Development, Population Policy, and Demographic Transition in the Republic of Korea*

106. David C. Cole and Yung Chul Park, *Financial Development in Korea, 1945–1978*

107. Roy Bahl, Chuk Kyo Kim, and Chong Kee Park, *Public Finances during the Korean Modernization Process*

94. Parks M. Coble, *The Shanghai Capitalists and the Nationalist Government, 1927–1937*

95. Noriko Kamachi, *Reform in China: Huang Tsun-hsien and the Japanese Model*

96. Richard Wich, *Sino-Soviet Crisis Politics: A Study of Political Change and Communication*

97. Lillian M. Li, *China's Silk Trade: Traditional Industry in the Modern World, 1842–1937*

98. R. David Arkush, *Fei Xiaotong and Sociology in Revolutionary China*

99. Kenneth Alan Grossberg, *Japan's Renaissance: The Politics of the Muromachi Bakufu*

100. James Reeve Pusey, *China and Charles Darwin*

101. Hoyt Cleveland Tillman, *Utilitarian Confucianism: Ch'en Liang's Challenge to Chu Hsi*

102. Thomas A. Stanley, *Ōsugi Sakae, Anarchist in Taishō Japan: The Creativity of the Ego*

103. Jonathan K. Ocko, *Bureaucratic Reform in Provincial China: Ting Jih-ch'ang in Restoration Kiangsu, 1867–1870*

104. James Reed, *The Missionary Mind and American East Asia Policy, 1911–1915*

105. Neil L. Waters, *Japan's Local Pragmatists: The Transition from Bakumatsu to Meiji in the Kawasaki Region*

108. William D. Wray, *Mitsubishi and the N.Y.K., 1870–1914: Business Strategy in the Japanese Shipping Industry*

109. Ralph William Huenemann, *The Dragon and the Iron Horse: The Economics of Railroads in China, 1876–1937*

110. Benjamin A. Elman, *From Philosophy to Philology: Intellectual and Social Aspects of Change in Late Imperial China*
111. Jane Kate Leonard, *Wei Yuan and China's Rediscovery of the Maritime World*
112. Luke S. K. Kwong, *A Mosaic of the Hundred Days: Personalities, Politics, and Ideas of 1898*
113. John E. Wills, Jr., *Embassies and Illusions: Dutch and Portuguese Envoys to K'ang-hsi, 1666–1687*
114. Joshua A. Fogel, *Politics and Sinology: The Case of Naitō Konan (1866–1934)*
115. Jeffrey C. Kinkley, ed., *After Mao: Chinese Literature and Society, 1978–1981*
116. C. Andrew Gerstle, *Circles of Fantasy: Convention in the Plays of Chikamatsu*
117. Andrew Gordon, *The Evolution of Labor Relations in Japan: Heavy Industry, 1853–1955*
118. Daniel K. Gardner, *Chu Hsi and the* Ta Hsueh: *Neo-Confucian Reflection on the Confucian Canon*
119. Christine Guth Kanda, *Shinzō: Hachiman Imagery and its Development*
120. Robert Borgen, *Sugawara no Michizane and the Early Heian Court*
121. Chang-tai Hung, *Going to the People: Chinese Intellectuals and Folk Literature, 1918–1937*
122. Michael A. Cusumano, *The Japanese Automobile Industry: Technology and Management at Nissan and Toyota*
124. Steven D. Carter, *The Road to Komatsubara: A Classical Reading of the Renga Hyakuin*
125. Katherine F. Bruner, John K. Fairbank, and Richard T. Smith, *Entering China's Service: Robert Hart's Journals, 1854–1863*

竹本内匠太夫

道頓堀蛭子橋より
東を見る圖
吉田